Computing and Systems Technology Division 2015

Core Programming Area at the 2015 AIChE Spring Meeting and 11th Global Congress on Process Safety

Austin, Texas, USA
26-30 April 2015

ISBN: 978-1-5108-0700-6

Printed from e-media with permission by:

Curran Associates, Inc.
57 Morehouse Lane
Red Hook, NY 12571

Some format issues inherent in the e-media version may also appear in this print version.

Printed by Curran Associates, Inc. (2015)

AIChE
3 Park Avenue
New York, NY 10016-5991

Phone: (203) 702-7660
Fax: (203) 775-5177

www.aiche.org

Additional copies of this publication are available from:

Curran Associates, Inc.
57 Morehouse Lane
Red Hook, NY 12571 USA
Phone: 845-758-0400
Fax: 845-758-2634
Email: curran@proceedings.com
Web: www.proceedings.com

TABLE OF CONTENTS

409505 Chemical Engineering: Is This the Golden Age?

Abstract:

Schlanger will offer his insights on the current opportunities in the chemical process industries in his talk entitled "Chemical Engineering: Is This the Golden Age?" on Monday morning, April 27. At the conclusion of his lecture, Schlanger will receive the AGILE Award in recognition of his leadership at CEVA Group, a $9 billion logistics supplier to industry worldwide; LyondellBasell, a major player in the plastics, chemicals, and refining industries; Arco Chemical; and at the other companies where he has held leadership positions.

399390 Big Data: What It Is, and Why You Should Care

Abstract:

Big Data: Surely one of the most confusing terms ever invented by the IT industry. ARC Advisory's data clearly reflects that confusion – 38% of manufacturers don't understand what Big Data is, or why they should care.

But, ready or not, Big Data is real - and it's coming to a plant near you. This presentation will:

- Define, in practical terms, what Big Data is. And, just as importantly, what it isn't...
- Showcase emerging examples of industrial Big Data applications
- Highlight key technologies, skills and processes needed for Big Data applications
- Provide actionable steps that all organizations can take to start gaining value from Big Data

Harness Big Data with Big Analytics

Keith R. Holdaway
SAS Institute Inc.

15 AIChE Spring Meeting, Austin, TX

sas

Agenda

- The Analytics Lifecycle
- The Intelligent Field
- Data Mining Virtuous Cycle
- Data Mining: What is it?
- Big Data: What is it?
- Subsurface Systems
- Analytical Methodologies
- Deterministic and Probabilistic
- SEMMA Process: Case Studies

"THAT'S your Ark for the Big Data flood? Noah, you will need a lot more storage space!"

15 AIChE Spring Meeting, Austin, TX

sas

The Analytics Lifecycle

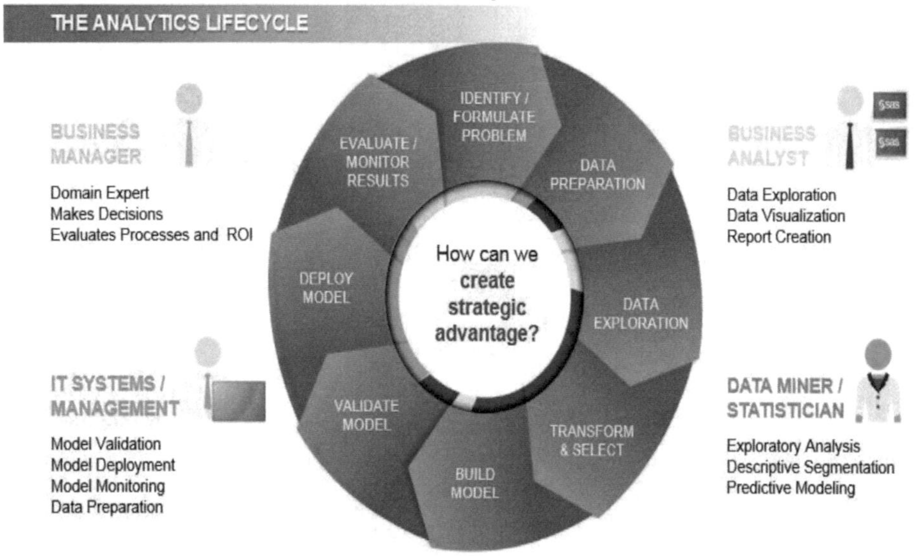

15 AIChE Spring Meeting, Austin, TX

sas

The Intelligent Field

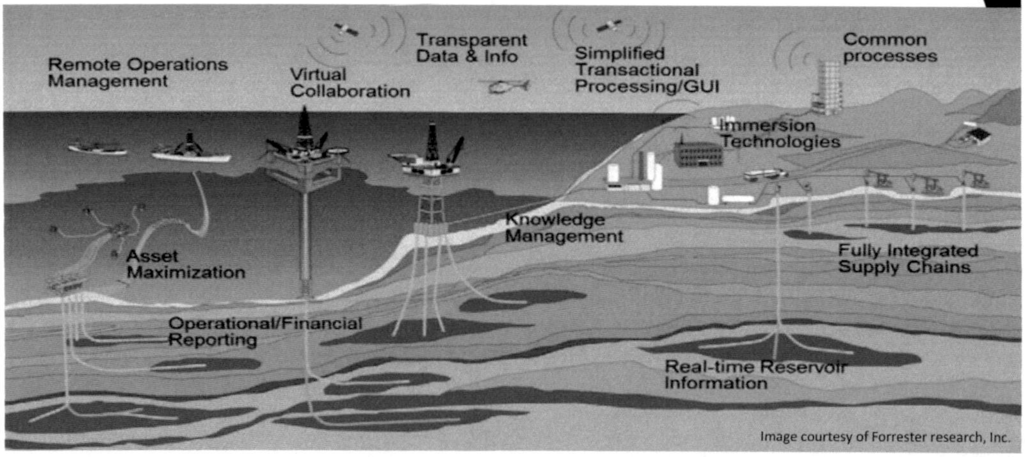

Image courtesy of Forrester research, Inc.

15 AIChE Spring Meeting, Austin, TX

sas

The Evolution of the Intelligent Field

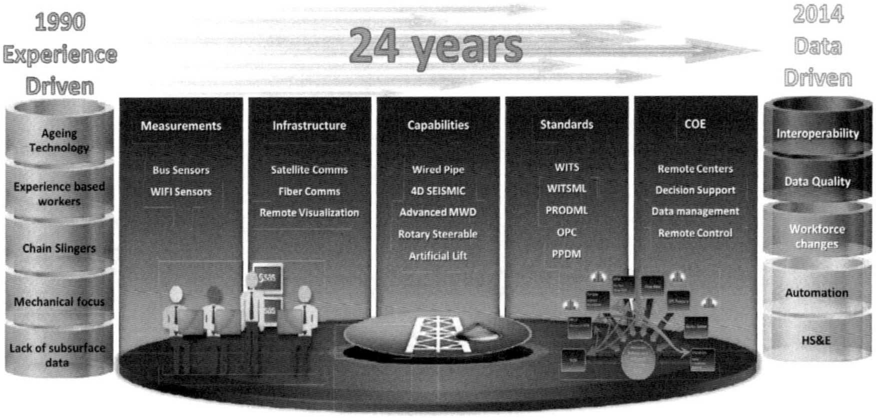

15 AIChE
Spring Meeting, Austin, TX

sas

Data Mining Virtuous Cycle

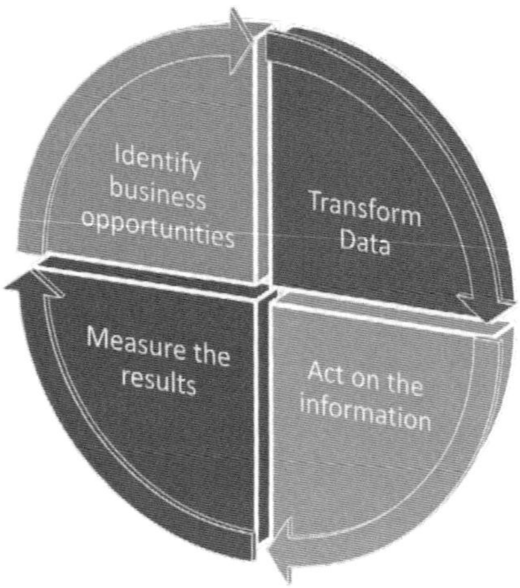

"Those who do not learn from the past are condemned to repeat it."

George Santayana

15 AIChE
Spring Meeting, Austin, TX

sas

Data Mining: What is it?

- Data Mining Styles
  - Hypothesis Testing
  - Directed Data Mining
  - Undirected Data Mining

15 AIChE Spring Meeting, Austin, TX

sas

Big Data: What is it?

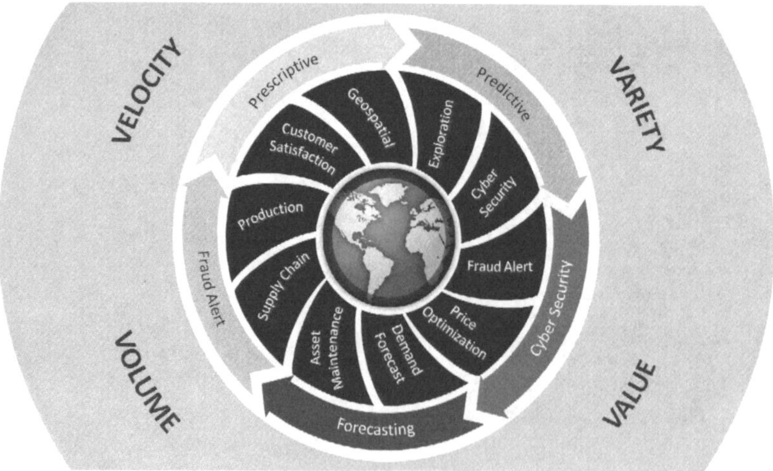

15 AIChE Spring Meeting, Austin, TX

sas

Subsurface Systems

COMPLEX DATA: OIL AND GAS

- Multivariant
- Multidimensional
- Multivariate
- Stochastic

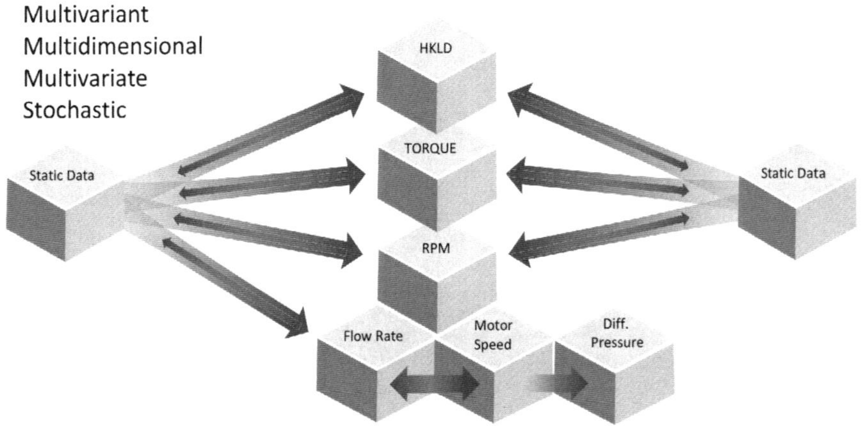

15 AIChE
Spring Meeting, Austin, TX

sas

Analytical methodologies

BATCH AND REAL-TIME INTEGRATED

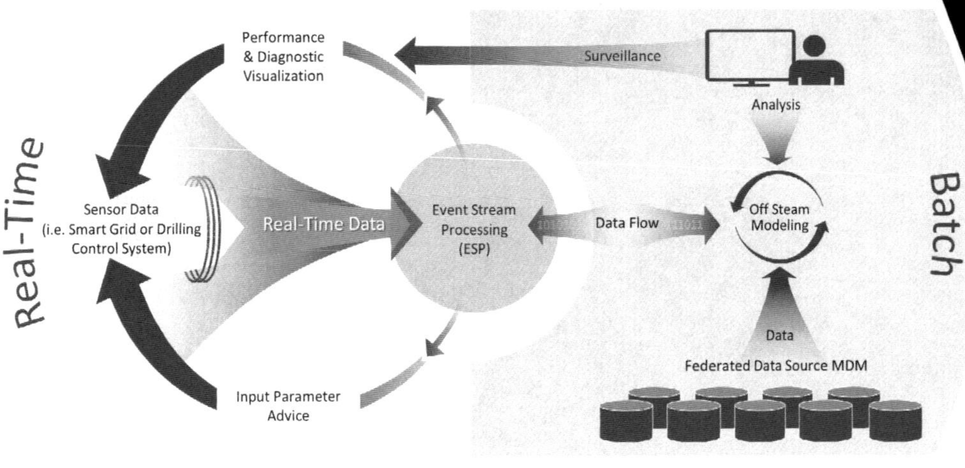

15 AIChE
Spring Meeting, Austin, TX

sas

Deterministic and Probabilistic

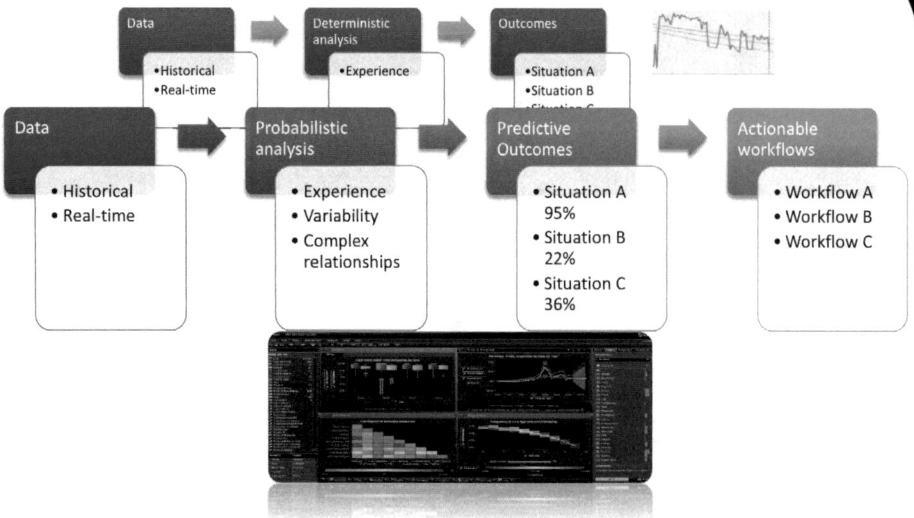

15 AIChE
Spring Meeting, Austin, TX

sas

SEMMA Process

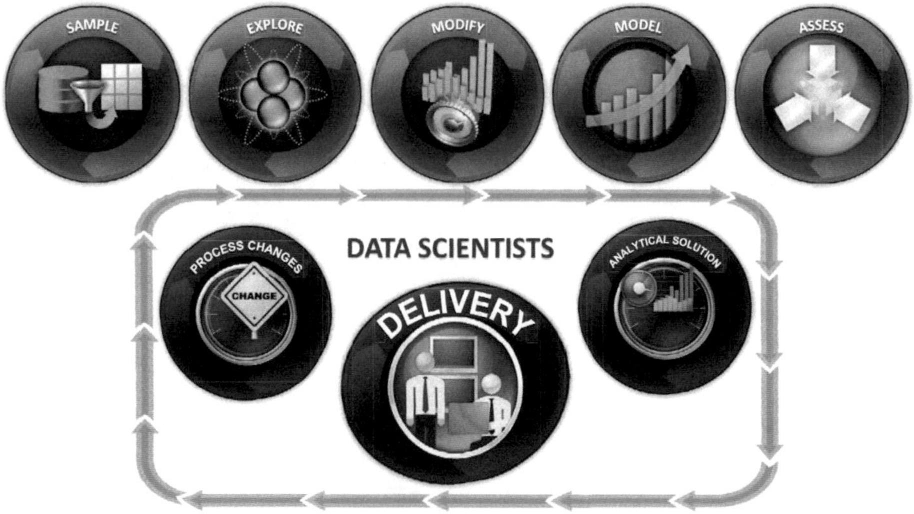

15 AIChE
Spring Meeting, Austin, TX

sas

Reserve Estimation and Production

For the oil and gas industry, production forecasting is a highly complex task, and as such requires an advanced analytical engine to achieve robust forecasts.

Jill Feblowitz
IDC Vice President
Utilities and Oil and Gas

BUSINESS ISSUES

Performing accurate analysis and interpretation of reservoir behavior is fundamental to assessing extant reserves and potential forecasts for production.

SOLUTION

Predictive analytics, combined with decline curve methodologies, provides more informative forecasting results of future production.
Access well and reservoir data quickly and easily
Apply analytical functions consistently
Perform decline analysis quickly
Estimate unconventional sources of well production more accurately

Results

Better availability and confidence in results.
Fewer data exceptions and more frequent, consistent reports.
More time for strategic planning, less time spent cleansing data
Engineers can evaluate multiple what-if scenarios for reservoir forecasts.

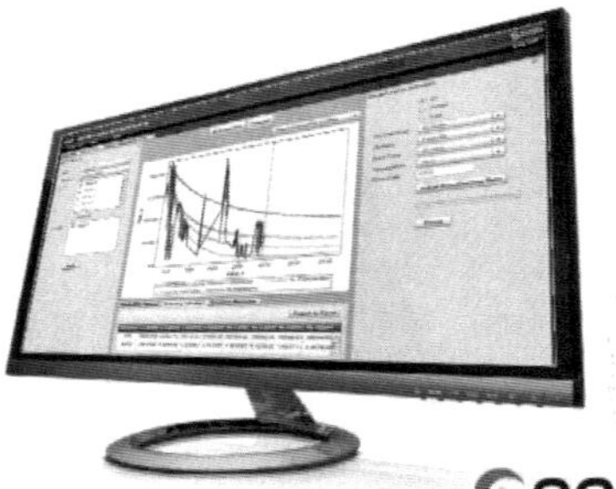

15 AIChE
Spring Meeting, Austin, TX

sas

Electrical Submersible Pumps

"Implementation of SAS Predictive Asset Maintenance offers us a competitive advantage by predicting and explaining performance anomalies in ultra-deep water production lift systems"

Remote Surveillance Project Manager

BUSINESS ISSUES

Electrical submersible pumps provide artificial lift to enhance hydrocarbon recovery. When a pump fails the operator faces the cost of deferred production while a replacement is mobilized as well as the replacement costs of the pump itself.

SOLUTION

Using a combination of Asset Performance Analytics and Event Stream Processing we were able to provide the customer with Real-time surveillance of motor frequency, prediction of time to fail and visualization of 10,000 ESP's across North and South America

RESULTS AND EXPECTED RESULTS

Real-time Proactive prediction of failure
3 months lead time on predicted failures
Downtime reduced to 6 hours
Estimated savings of $3 million on a Deepwater installation

15 AIChE
Spring Meeting, Austin, TX

sas

BUSINESS ISSUES - SPE-135523-MS
Multinational operator was trying to qualify well performance in the Pinedale Anticline in western Wyoming. The environment presented several challenges because of commingled productivity from multiple fluvial sand packages in a single wellbore distributed over greater than 5000 feet of gross vertical section

SOLUTION
Generation of a neural network that qualified the relationships between Petrophysical, Geological and Operational Parameters such that the solution could be used in both design and operational phases.

RESULTS AND EXPECTED RESULTS
Data Driven model that assisted in design and operation of

- Well placement
- Stage management
- Completions design
- Proppant design

Unconventional Gas

"SAS Enterprise miner enabled us to accelerate and implement a easy to use and intuitive solution to the multivariate uncertainties inherent in subsurface environments"

Production Engineering Advisor

15 AIChE
Spring Meeting, Austin, TX

sas

Business Issue

- Increase operational lifetime of low-volume, steam-injected wells
- Maximize production output by improving equipment reliability
- Reduce operational costs and optimize maintenance schedules
- Develop a culture of analytics for more informed decisions

Solution
SAS® provides asset failure predictions, real-time surveillance and descriptive visual analytics

Expected Results

- Lower costs through more efficient product and maintenance processes
- Potential to increase production at the rate of $200M in incremental annual revenue
- Longer production life of mature, resource-intensive wells
- Share insights from real-time surveillance and predictive analytics on mobile devices

US Energy Exploration Company

"We will monitor and optimize all assets 24 hours a day, seven days a week, 365 days a year."

"Producing an extra one quarter to half a barrel per well each day will provide as much as $200 million a year in incremental revenue."

15 AIChE
Spring Meeting, Austin, TX

sas

Production Optimization in Brownfield Strategies

BUSINESS ISSUES

Multinational operator was trying to identify key performance indicators in mature fields with a relatively long production history. The assets showed a rapid decrease in pressure; degradation of GOR and Water cut in each well. From a business perspective, the operator anticipated an accelerated decrease in anticipated profitability from improvement activities – NPV from new capital projects (new wells) and margins from OPEX (work over).

SOLUTION

Development of a suite of analytical workflows that generated a 2D data set from Principal Component Analysis, Multivariate Analysis and clustering of a multidimensional input space across space and time.

RESULTS AND EXPECTED RESULTS

Data Driven model that enabled more effective strategies and tactics for:

- New well locations
- Well production optimization
- Managing and predicting water cuts and GOR
- Depletion and pressure maintenance

"Multivariate analysis and cluster analysis classifies the well portfolio to identify key production indicators. This allowed us to segment the field and identify poorly-drained zones"

Jean-Paul Valois, Total

15 AIChE Spring Meeting, Austin, TX

sas

Q&A

15 AIChE Spring Meeting, Austin, TX

sas

401987 Smart Grid: The Basics - What? Why? Who? and How?

Abstract:

The electric power industry is poised for an unprecedented change in the way it operates. Motivated by forecasted changes in its generation profile, including a greater use of renewable sources, the power industry is looking to create a more interactive relationship with its consumers. Since the industrial community consumes about one third of all generated power, this sector is expected to play a major role in the transition. To help decipher the upcoming changes we will attempt to answer the following questions: What is a smart grid? Why is it being created? Who stands to benefit? and How does one participate?

402071 Smart Grid: You Are Not Alone – the Role of Curtailment Service Providers

Abstract:

As the grid becomes "smarter," the chemical process industries (CPI) are increasingly finding opportunities to reduce overall power costs, and in many cases, develop new revenue streams. However, finding these new opportunities within the rapidly changing power industry can be a daunting task. Fortunately, third-party consultants known as curtailment service providers (CSP) can help you navigate this ever-changing landscape. This presentation takes a high-level look at industrial demand-response (DR) opportunities in the U.S. deregulated power markets, the revenue potential of these opportunities, and the process for implementing a smart-grid opportunity at a CPI facility.

INDUSTRIAL BIG DATA VISION AND SOLUTIONS

Mark Nixon, Terry Blevins and Willy Wojsznis
Innovation Center, Emerson Process Management, Austin, TX

Extended Abstract

Big Data is a ground breaking technology for many industries. Every industry is faced with the challenge of how to implement technology to achieve maximum benefits.

The process industries adopt many Big Data approaches that are applied in other industries however the Big Data implementation for Process Industries is distinctive in that it sets specific requirements for Big Data infrastructure, learning algorithms including data analytics, and presenting the results.

The process industry's Big Data infrastructure uses networking and database structures that have many common features with existing applications. In addition the infrastructure also includes structures such as lab data, NIR data, and spectral data; items that are not normally incorporated into control systems. Big Data treats all of this normally unincorporated data as unstructured. Schema information is combined with unstructured data to allow a wide range of analytics to be performed both off-line and on-line.

Process industry data includes interdependent time series data that requires efficient streaming, storage, and call-up. For meaningful analytics to be performed, real-time analysis must be combined with data context. Calculations are performed through the stream processing and calculation engine. Common Big Data techniques applied to the process industries require extensive adaptation or redevelopment, typically outside of the control system itself.

The presentation will address the basic components of Big Data pipeline for the process industry: hardware and software infrastructure, data streaming, data preprocessing and data learning techniques.

The core of data learning is Data Analytics (DA) which has proven its effectiveness in process fault detection and quality prediction both for batch and continuous processes. The real prospects are that Big Data based on DA will be among the leading directions for improving process effectiveness. DA requires a significant departure from the traditional thinking about how process control is implemented. Instead of the deterministic and tangible world of signals and devices, there is an abstract realm of statistical indexes, correlation factors and matrix operations. This puts a significant strain on the control systems' developers, engineering companies, process operation and maintenance personnel. The presentation will address these major challenges for professionals working on Big Data for the process industry.

References

1. Peter Harrington *Machine Learning in Action*, Manning, 2012
2. Sean Owen, Robert Anil, Ted Dunning and Ellen Friedman *Manhout in Action,* Manning, 2012
3. Christopher M. Bishop *Pettern Recognition and Machine Learning,* Springer, 2009
4. Terrence Blevins, Willy K. Wojsznis and Mark Nixon *Advanced Control Foundation,* ISA, 2013
5. Dunia, R., Edgar, T., Blevins, T., Wojsznis, W., *Multistate Analytics for Continuous Processes*, Journal of Process Control, 2012
6. Kourti, T. *Application of latent variable methods to process control and multivariate statistical process control in industry*. International Journal of Adaptive Control and Signal Processing 2005; 19:213-246.

401870 Big Data

Abstract:

Today the term big data draws a lot of attention, but behind the hype there's a simple story. For decades, companies have been making business decisions based on transactional data stored in relational databases. Beyond that critical data, however, is a potential treasure trove of non-traditional, less structured data: weblogs, social media, email, sensors, and photographs that can be mined for useful information. Decreases in the cost of both storage and compute power have made it feasible to collect this data - which would have been thrown away only a few years ago. As a result, more and more companies are looking to include non-traditional yet potentially very valuable data with their traditional enterprise data in their business intelligence analysis.

To derive real business value from big data, you need the right tools to capture and organize a wide variety of data types from different sources, and to be able to easily analyze it within the context of all your enterprise data. Oracle offers the broadest and most integrated portfolio of products to help you acquire and organize these diverse data types and analyze them alongside your existing data to find new insights and capitalize on hidden relationships.

SQL is the natural language for querying data, but data increasingly lives in many places: HDFS, HBase, relational databases, and NoSQL stores. As the options for data management explode, developers, applications, and tools begin to demand SQL everywhere.

In this session, we'll present a SQL-based strategy for integrating Hadoop, NoSQL technologies, and relational databases. Our approach illustrates how to get the best out of each platform, and enables integration with no changes to existing applications. Along the way, we'll dive deep into the architecture of Big Data SQL, which can access all of these platforms in a single query through query franchising

402072 Smart Grid: Communicating with the Smart Grid – the Role of Automated Demand Response Communications Systems

Abstract:

In 2002 researchers in California began to develop a low-cost demand response (DR) communication and automation system that would send electricity price and grid reliability signals and that could easily be connected to existing industrial control systems. The technology was called Open Automated Demand Response (OpenADR) — open to distinguish it from other, proprietary automated DR efforts. The system uses a client-server communication architecture with an open application-programming interface (API) and a web server that communicates with remote software clients. In OpenADR, the client side of DR automation is deployed via software clients embedded in control systems or deployed with gateways. The DR Automation Server (DRAS) publishes signals from the utility, grid operator, or DRAS operator to communicate with the remote client, and the client uses pull technology to request event data from the DRAS every minute. In 2009 the National Institute of Standards and Technology (NIST) lead the activity in the U.S. to identify, develop, and implement Smart Grid interoperability standards. OpenADR 2.0 was a result of a set of profile specifications developed from the formal standardization of OpenADR.

402073 Smart Grid: Industry Meets the Smart Grid – Integrating Process Automation and Energy Management in Industrial Facilities

Abstract:

A smart grid opens new cost saving and revenue generating possibilities for process automation. The flexibility provided by integrated process automation and energy management helps manufacturers not only to save energy by using electricity more efficiently, but also to procure it under the best possible financial terms. Demand response (DR) is one way of demand modification, i.e. the change in energy consumption pattern, in response to grid and electricity market conditions through voluntary market based mechanism. In direct load control, demand modification is carried out by the grid operator directly, in demand response, the system operators achieve demand modification indirectly through price signals. The DR program participants carry out those instructions and provide demand response by turning down or turning off some equipment or processes. In automated demand response, the ISO or the utility manipulates demand directly, which is referred to as direct load control (DLC). This is an established technology for which optimization-based automation solutions already exist, and it is currently being used by many utilities, for example to turn off or cycle large appliances such as air conditioning systems and hot water heaters. Another important incentive comes from adapting the electricity load according to day-ahead pricing. In this case the production profile is determined by scheduling the production based on electricity price forecasts. This approach can create stability both on the utility and the manufacturing sites and therefore enable significant savings. The focus here is on similar capabilities for large consumers within the chemical process industries (CPI)

Importance of Modeling and Simulation in Framing a Process Development Career

Joseph B. Powell, Ph.D.
Shell Chief Scientist – Chemical Engineering

Overview

Process development is a key skill for industry, and the ability to develop and use economic, process, and specific technology component models is a key skill for execution of this role.

At the entry of the R&D portfolio funnel, new ideas and opportunities must be constantly screened and evaluated to select best options to purse. Screening is best accomplished via developing and comparing economic models of available options. In the absence of early information or data or performance, one can assume best possible yields and conversions, to determine which R&D options present the greatest opportunity if successfully developed. As the project progresses, the economic models are further refined to more accurately reflect required capital costs and actual yields, and to include full discounted cash flow assessments, earning power, and net present value for those projects which do progress to full development and deployment. The level of detail increases to allow a final investment decision to be made. Experienced process development engineers provide exceptional value to their organizations via the ability to provide preliminary cost estimates, odds for success, and identification of potential issues or show stoppers in early idea assessment, where limited information is available.

In Exploration and Production, modeling of enhanced oil recovery projects is the domain of supercomputers. However, a simplified model can be instructive in understanding impacts of the level of interfacial tension reduction, well spacings and capillary number, and mobility ratio of sweep efficiency, for use in process optimization. Reservoir heterogeneity often defeats the attempts of rigorous modeling via supercomputer, but does point to the fact that better characterization is needed to improve the accuracy of prediction.

Fixed-bed adsorption is another example where complex models resulting in stiff numerical equations and excessive computation and analysis time can be used, and might be the initial method of choice. However, responses typically fall into self sharpening or dispersive front behavior [1,2], such that analytical solutions for either of these limits may provide more than enough basis for commercial design, relative to the accuracy of system parameters such as mass transfer or dispersion coefficients. In the end, engineering safety factors in design may likely eliminate the need for rigorous calculation, if transport parameters were known with the accuracy needed to enable a rigorous simulation. Considerable time may be saved by recognizing these limits, and using models that are fit-for-purpose.

Pressure drop through a bed of compressible solids is another example where modeling enabled the use of model fluids for experimentation, and provided a more reliable result than direct pilot testing. Ability to develop and apply a model (modified Ergun equation) with parameters derived from small scale laboratory testing, enabled prediction of scale effects as a function of applied stress, which otherwise could not be seen without full scale testing of deployment concepts. Catastrophic failure upon scale-up from pilot to commercial was thus be averted.

Other examples include modeling of distillation via reflux experiments to obtain separation factors for use in design for real, complex mixtures, development of a commercial reactor model (hydroformylation) used to control reactor temperature profiles to avoid catastrophic loss of catalyst to extraction, development of a liquid extractor model to show the basis for counter intuitive response needed to prevent loss of interface level (via impact on slip velocity), development of effectiveness factor models to show the dramatic impact of water as solvent as a function of reactant concentration, in limiting hydrogen solubility, leading to starvation and catalyst deactivation at high volumetric reaction rates. Additional examples include adaptation of kinetics to enable substitution of a lower stench, lower volatility reaction promoter to enable low pressure reaction experiments to be conducted safely in glassware for modified catalyst development; modeling of heat losses in small scale lab reactor to project harmful runway if translated to more adiabatic, large scale commercial units; understanding that entrainment at the top of a reactor scales with linear velocity whereas reactor conversion scales with residence time, to enable high velocity testing for entrainment risk via transient operation of pilot equipment; scaling down reaction kinetics to enable poorly stirred multi throughput screening reactors to be effective in diagnosing differences in catalyst performance (vs. study of gas-liquid mass transfer), development of a transient CSTR model to enable manipulation of long-term catalyst life test units to test for catalyst activity above statistical noise level of steady state response.

Key aspects of many of these model development examples is the ability to develop fit-for-purpose, simplified or limiting-form models of key steps or principles, to generate results which are intuitively understandable and may be rapidly used to test alternate scenarios, vs. attempting rigorous modeling of all transport, rate, and equilibria, in a more complex model. The latter can allow scenarios and cases to be generated and may be warranted for some design or optimization problems, but may require considerable time and resource to develop, and underlying cause-and-effect relationships may not be readily evident. Where undertaken, results of rigorous and more comprehensive modeling should be tested via the logic and understanding of a simpler first principles model, to avoid risk of gross parameter error.

Effective use of modeling is a short cut to higher productivity in problem solving, while also providing opportunity for continued professional learning. Ability to reduce a problem to a first-principles model and provide limiting form solutions is an effective way to rapidly screen options, avoid expensive "black box testing" at larger scale, and insure accuracy of final designs.

References

1. Vermeulen, T., M. D. LeVan, N. K. Hiester, and G. Klein, *Adsorption and Ion Exchange,* Section 16 in "**Perry's Chemical Engineers Handbook**", sixth ed., R.H. Perry, D.W. Green, and J.O. Maloney, eds. McGraw-Hill, New York (1984).
2. Rhee, H.-K., B. Bodin, N. Amundson, *A study of the shock layer in equilibrium exchange systems,* **Chem. Eng. Sci.**, **26**, 1571–1580 (1971).

Classical PID control in presence of missing data using compressed sensing techniques

Satheesh K. Perepu and Arun K. Tangirala*
Dept. of Chemical Engineering, IIT Madras, Chennai 600036, India
Email:satheesh841@gmail.com and arunkt@iitm.ac.in

***Abstract*—Most of the controllers used in process industries are proportional, integral and derivative (PID) controllers. Missing data is common in these industries due to many reasons such as sensor malfunctioning, data transmission loss *etc*. The performance of PID control scheme will degrade if operated in presence of missing data. Hence in this paper a method based on compressed sensing techniques is proposed for online reconstruction of missing data and performance is shown with an application to process control. The proposed algorithm is demonstrated on three simulated studies and results show that the proposed algorithm gives comparable results with classical PID control scheme with no missing data.**

Keywords*—*process control;compressed sensing;random sampling;MAR;PI control;

I. Introduction

Maintaining the process variable at a specified level (setpoint) is one of the important task in all process industries. This task can be challenging because the controller has to achieve this in presence of external disturbances. These controllers can be of basic type such as PID or of advance type such as model predictive controllers (MPC). Almost 95% of closed loop controllers used in industries are of PID type [18]. The PID controllers are most popular in industries because of their ease of maintenance and tuning. There are some hundreds of tuning rules available in literature to tune PID controllers [13]. The tuning rule used in the paper is Ziegler-Nichols open loop tuning rule which is proposed by Ziegler [19]. He developed this tuning rules based on peak sensitivity minimisation.

An important assumption made in all classical PID control schemes is that sensor measurements are available without fail at each sampling instant. However, it is not uncommon to encounter situations involving sensor and/or communication failures leading to the case of *missing samples*. Clearly such situations present several challenges to control and other operations such as monitoring and identification. In such situations a process model based on first principles will be handy. Although they provide accurate description of the process, they have two primary disadvantages (i) they require full knowledge of parameters affecting the system which is difficult to obtain (ii) solving ordinary differential equations which can be solved using numerical techniques. To address these difficulties data driven models have been developed [11], which simplify the modelling by implementing simple heuristic relationships between input and output of the system. At the same time, these data-driven models also have disadvantages: process dynamics are not fully captured by data-driven models; in addition, different data-driven models are based on different assumptions, and there is a lack of clear understanding on which assumption agrees with physical process better. Hence a method which estimates missing data online and doesn't require an explicit model is needed.

Developing methods for missing data reconstruction has been a hot topic of research for over a century now [15]. Several researchers have addressed the problem of missing samples in different applications, especially in identification, using some interesting statistical and optimal estimation approaches. However these techniques are computationally expensive restricting them to offline applications [1, 12]. To the best knowledge of authors, the problem of online missing data recovery, especially in classical control, has not been sufficiently addressed. Hence there is a need to develop a method which estimates missing data online. The main requirement of such a method is that the time needed for computing missing data should be less than that of the sampling interval chosen for the process.

In this work, a method directed at online missing data recovery for control using ideas of compressed sensing (CS) [4] is proposed. The goal in compressed sensing is to recover $\mathbf{x}$ by first estimating the sparse representation $\mathbf{s}$ given $\mathbf{A}$ or simultaneously along with $\mathbf{A}$ when it is unknown [14]. In general there will be many features in a signal like steady state, oscillatory, transients etc. Hence to find a single basis in which the entire signal has same degree of sparsity (same number of non-zeros) is difficult. Hence an idea of constructing composite dictionary [5] is used in the paper. The proposed method is discussed in §III. The two prime features of the proposed method are that, (i) it is computationally efficient and (ii) does not involve any *explicit* process model.

The rest of the paper is organized as follows. The necessary essentials of compressed sensing are reviewed in Section II. Section III presents the proposed algorithm for estimation of missing data online. Results from simulation studies are presented in Section IV. The paper closes with a few concluding remarks in Section V.

II. Foundations of Compressed Sensing

The central concept in compressed sensing (CS) is that of *sparse representation*, which refers to a minimal set of non-zero coefficients that represent the signal in some basis space. One of the main problems of interest in the compressed sensing literature is to recover N observations of a signal $z[k]$ from its M random measurements ($M \ll N$), call it $s[k]$, under the premise that $z[k]$ is K-sparse (number of non-zeros) ($K < M$) in some basis space [4]. The measurement vector $s[k]$ can be in the same domain as that of $z[k]$, for e.g., randomly subsampled

$z[k]$, or in some other transformed domain. We shall very briefly review the basic mathematics underlying these ideas below.

$$\mathbf{s} = \Pi\mathbf{z}, \quad \text{and} \quad \mathbf{z} = \mathbf{B}\mathbf{x} \tag{1}$$

$$\Longrightarrow \mathbf{s} = \mathbf{A}\mathbf{x} \tag{2}$$

where Π is $M \times N$ measurement matrix (when $\mathbf{s}$ is subsampled $\mathbf{z}$, it consists of ones and zeros), $\mathbf{x} \in \mathbb{R}^{N\times 1}$ is the sparse representation of $\mathbf{z}$ in some *sparsifying* basis space defined by the matrix $\mathbf{B}$. In (2), the matrix $\mathbf{A} = \Pi\mathbf{B}$ is said to be the *dictionary* for $\mathbf{s}$, which may be known a priori (fixed) or unknown (adaptive) [14]. The goal in compressed sensing is to recover $\mathbf{z}$ by first estimating the sparse representation $\mathbf{x}$ given $\mathbf{A}$ or simultaneously along with $\mathbf{A}$ when it is unknown .

An N-long vector $\mathbf{x}$ is said to be sparse if the number of non-zero elements in $\mathbf{x}$ is less than the number of zero-valued elements. There exist several measures of sparsity. See [10] for a comprehensive study of different sparsity measures. The zero-norm of a vector, given by

$$||\mathbf{x}||_0 = \dim\{i|x_i \neq 0\} \tag{3}$$

is frequently used in the CS literature as a measure of sparsity.

If the dictionary $\mathbf{A}$ is known a priori, the sparse representation $\mathbf{x}$ is recovered by solving the optimization problem

$$\min_{\mathbf{x}} ||\mathbf{x}||_0 \quad \text{subject to} \quad \mathbf{s} = \mathbf{A}\mathbf{x} \tag{4}$$

However, the problem is non-convex and NP-hard. A widely used alternative is to replace with a convex 1-norm optimization to solve (2), as put forth by [4]:

$$\min_{\mathbf{x}} ||\mathbf{x}||_1 \quad \text{subject to} \quad \mathbf{s} = \mathbf{A}\mathbf{x} \tag{5}$$

Candes et. al. [2] proved that the zero-norm solution equals 1-norm solution as long as $\mathbf{A}$ satisfies what is known as the Restrictive Isometric Property (RIP), given by

$$(1-\delta)\,||\mathbf{x}||_2^2 \leq ||\mathbf{A}\mathbf{x}||_2^2 \leq (1+\delta)\,||\mathbf{x}||_2^2 \tag{6}$$

It is difficult to verify RIP property in practice. Hence, Haupt et. al. [9] proved that the Toeplitz matrices (quasi to be specific) satisfy RIP theoretically. Elad [6] replaced RIP with a weaker condition based on *mutual coherence*. The mutual coherence of matrix $\mathbf{U} = [\mathbf{u}_1 \quad \mathbf{u}_2 \quad \cdots \quad \mathbf{u}_p]$ is given by

$$\mu(\mathbf{U}) = \frac{1}{p}\max_{i\neq j}\frac{|\mathbf{u}_i^T\,\mathbf{u}_j|}{||\mathbf{u}_i||_2\,||\mathbf{u}_j||_2} \tag{7}$$

where $||.||_2$ is the standard 2-norm. The mutual coherence of any matrix always satisfies $1/\sqrt{p} \leq \mu(\mathbf{U}) \leq 1$; see [16]. In order to guarantee recovery of the sparse vector, the mutual coherence of the dictionary $\mathbf{A}$ should be as low as possible [6]. Tropp [17] gives a condition for the exact recovery of $\mathbf{s}$ in terms of the support vector of $\mathbf{A}$. The precise requirement is that at least K columns of $\mathbf{A}$ be orthogonal.

There exist two classes of algorithms to solve (5) [6]. One class is that of greedy algorithms and other comprises iterative algorithms. Greedy algorithms include orthogonal matching pursuit (OMP), block-OMP, *etc*, while the iterative algorithms include least absolute shrinkage and selection operator (LASSO), basis pursuit (BP), basis pursuit de noising (BPDN), *etc*. Among these the widely used techniques to solve (5) are the LASSO and BPDN. We choose to employ the BPDN in this work due to its robustness (to noise) property. Another advantage is that even when $\mathbf{x}$ is not sufficiently sparse, $\mathbf{s}$ can be approximated with a small error. In BPDN (5) is solved as

$$\min_{\mathbf{x}} ||\mathbf{x}||_1 \quad \text{such that} \quad ||\mathbf{s} - \mathbf{A}\mathbf{x}||_2 \leq \zeta \tag{8}$$

where ζ is a small non-zero value. There are several flavours of the BPDN algorithm. The spectral gradient-based BPDN (SPG-BPDN) is used in this work.

As described in §I the signal can have many features which are sparse in different basis. For example suppose some part of the signal is sparse in dictionary $\mathbf{C}$ and some part in dictionary $\mathbf{D}$. In this case matrix $\mathbf{B}$ given in (2) is constructed by concatenating two matrices $\mathbf{C}$ and $\mathbf{D}$ i.e.

$$\mathbf{B} = [\mathbf{C} \quad \mathbf{D}] \tag{9}$$

The idea of constructing composite dictionary was proposed by [5] and they used it to reduce noise from the image. Later it was used by [8] to estimate missing data in sparse networks. In the current work we used the idea of composite dictionary to estimate missing data online and shown the performance of the method with an application to process control.

III. Proposed Algorithm

The proposed algorithm to estimate missing data online is explained as follows. Let $y[k]$ be the measured output at the kth sampling instant. Assume first N (typically very small) samples are available without any missing data. Then, the $(N+1)^{\text{th}}$ sample is predicted (without the aid of an *explicit* model) using the proposed algorithm. If the $(N+1)^{\text{th}}$ sample is missing, the predicted (estimated) value is passed on to the controller. On the other hand, when the $(N+1)^{th}$ sample is available, it is directly used to calculate the control action. It is assumed that the missing data mechanism is of *missing at random* (MAR) type.

The $(N+1)^{th}$ sample is computed using the following procedure. First, set up the equations

$$\underbrace{\begin{bmatrix} y[2] \\ \vdots \\ y[N+1] \end{bmatrix}}_{\mathbf{s}} = \underbrace{\begin{bmatrix} y[1] & \\ \vdots & \mathbf{B} \\ y[N] & \end{bmatrix}}_{\mathbf{A}} \mathbf{x} \tag{10}$$

where $\mathbf{B}$ is an $N \times N$ matrix of discrete Fourier transform (DFT) bases and $\mathbf{x}$ is the sparse representation of $\mathbf{s}$ in the overcomplete dictionary $\mathbf{A}$; see [7]. Note that the size of $\mathbf{A}$ is $N \times (N+1)$. The set of equations in (10) is essentially a sparse representation of the output in a *composite* basis. A useful viewpoint is that the past observations in $\mathbf{A}$ capture the dynamics of the process (as auto-regressive), while the DFT basis accounts for the steady-state/oscillatory characteristics. In (10), $y[N+1]$ and $\mathbf{x}$ are unknowns. Therefore, $\mathbf{x}$ is first estimated by leaving out $y[N+1]$ from (10),

$$\mathbf{s}(1:N-1) = \mathbf{A}(1:N-1,:)\mathbf{x} \tag{11}$$

where $\mathbf{A}(1:N-1,:)$ refers to the first $N-1$ rows of $\mathbf{A}$ and likewise for $\mathbf{s}(1:N-1)$. A standard CS algorithm such as the basis pursuit de-noising (BPDN) is used to estimate $\mathbf{x}$ in (11). Subsequently, $y[N+1]$ is estimated (predicted) as

$$\hat{y}[N+1] = \mathbf{A}(N,:)\mathbf{x}$$

To calculate $y[N+2]$, the output vector $\mathbf{s}$ and the dictionary $\mathbf{A}$ are updated as

$$\mathbf{s} = \begin{bmatrix} y[3] \\ \vdots \\ y[N+2] \end{bmatrix}, \quad \mathbf{A} = \begin{bmatrix} y[2] & y[1] & \\ \vdots & \vdots & \mathbf{B} \\ y[N+1] & y[N] & \end{bmatrix}$$

and the above procedure is repeated. This is continued for subsequent time instants. In order to prevent the growth of $\mathbf{A}$ in size, the number of columns containing past observations may be restricted sufficiently. The value of N is chosen based on dynamics of the system. Systems with slow dynamics, i.e., large time constants (relative to sampling time), require large N, while faster dynamics call for smaller block sizes.

An important requirement of close loop controller should possess is that the algorithm should not destabilize the closed loop system. In the proposed algorithm the stability of the resulting closed-loop system is supported by the fact that $\mathbf{x}$ can be reconstructed uniquely with high probability, provided the size of $\mathbf{s}$ is $K \log N$, where K is number of non-zero elements in $\mathbf{x}$; see [4]. Its tough to prove convergence of the algorithm theoretically since the proposed one involves solving 1 norm minimization problem which doesn't have an analytical solution. Hence, the stability is proved in this paper using Monte-Carlo simulations and is discussed in §IV.

IV. Simulated Case Studies

The proposed algorithm is demonstrated on three simulated case studies. In the first two case studies, we consider both the servo and regulatory problems. In third and final case study, we consider only the servo problem. A PI controller is used in both studies and the data is assumed to be MAR, with 80% of missing data. The controllers are tuned using the standard Ziegler-Nichols closed loop tuning rules.

A. *Case Study 1*

The system and controller under consideration are

$$G_p(s) = \frac{1}{5s+1} \tag{12}$$

$$G_c(s) = \frac{s+1}{s} \tag{13}$$

Sampling time of the process is $T_s = 0.5$ sec and hence the value of N chosen as 10 samples. To make simulation realistic, Gaussian white noise is added to the data. Variance of white noise is calculated such that the value of SNR is 10. In this example, the percentage of missing data is coonsidered as 80%. The profile of process variable for both servo and regulatory cases are shown in Figures 1 and 2 respectively. In both the plots, the solid lines correspond to process variable, and the locations where samples are available are marked as per the legend. From the plots, it is evident that the controller performs well even in presence of high percentage of missing data. It took 0.07 sec to estimate the missing sample for the proposed algorithm which is negligible when compared to the sampling time chosen for the process. Thus the prime requirement of the algorithm is satisfied. In the case of regulatory response, at least one sample should be available during the time when the effect of disturbance seen on the output of the system. In reality this seems to be a limitation for the proposed algorithm, but the probability of having no sample during the interval the effect of disturbance is seen on the output is low.

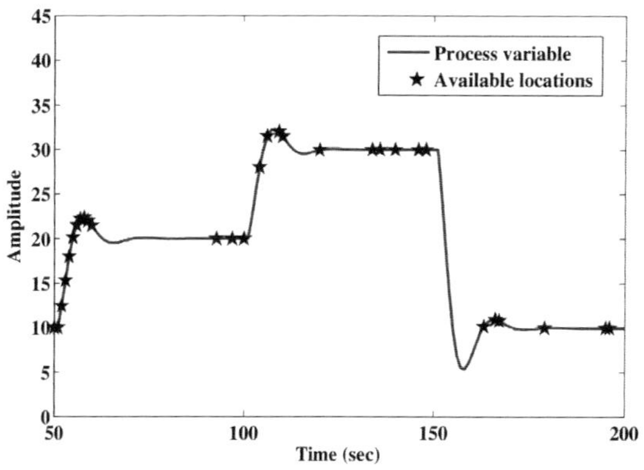

Fig. 1. First order system with PI control

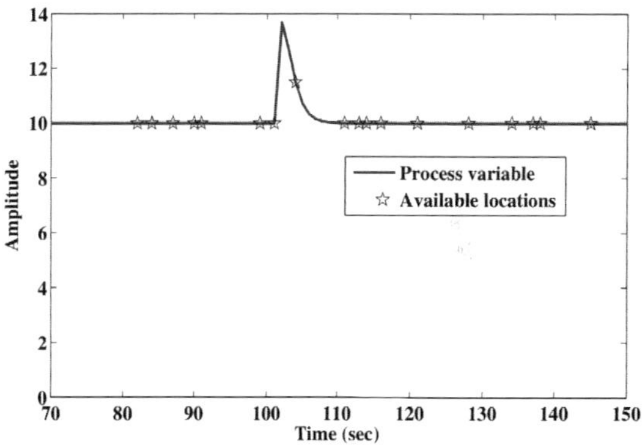

Fig. 2. First order system with regulatory control

B. *Case Study 2*

In this case study, the system chosen is a second order system which is given by

$$G_p(s) = \frac{1}{s^2+2s+1}$$

The controller designed using Ziegler-Nichols tuning rules and is obtained as

$$G_c(s) = \frac{0.2s+1}{s}$$

The sampling time is chosen as $T_s = 0.2$ sec and the value of N chosen as 5 samples. Measurement noise is introduced such that the SNR is 10. Plots of the controlled output for the servo and regulatory cases are shown in Figure 3 and 4, respectively. The solid line in Figures 3, 4 is the output profile including both measurements and estimated data (wherever measurements are missing). The available data are marked as indicated in the legend. From the plots it is evident that the controller performs well even in presence of high percentage of missing data. The computational time required to compute the

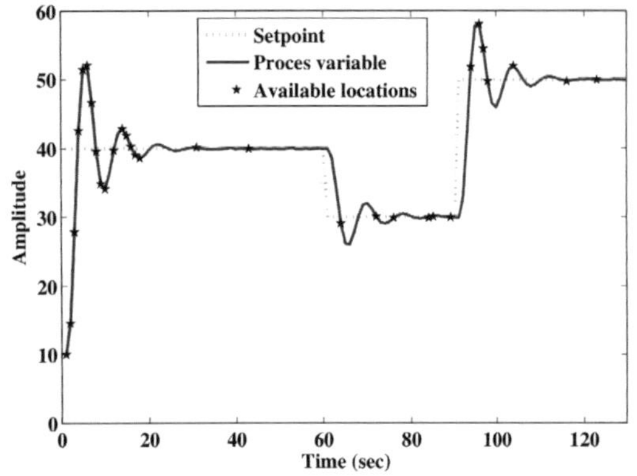

Fig. 3. Second order system with PI control

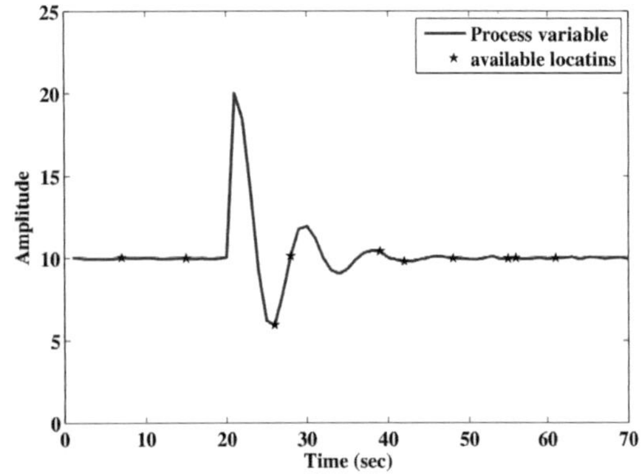

Fig. 4. Second order system with regulatory control

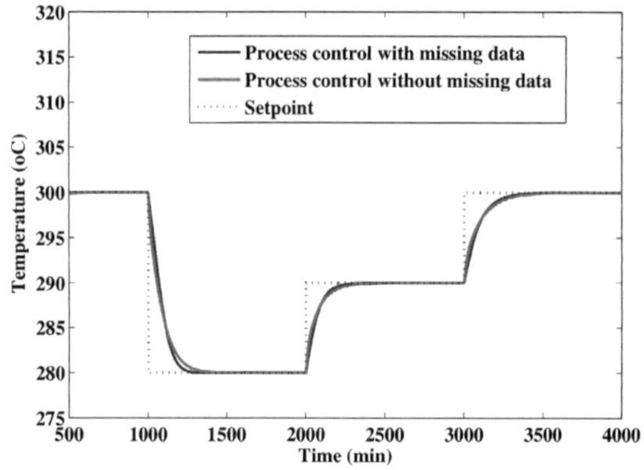

Fig. 5. CSTH with PI control

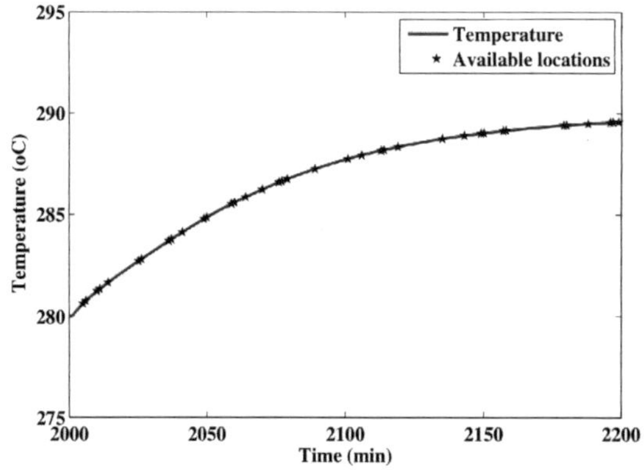

Fig. 6. Single setpoint change in temperature for CSTH

one-step ahead prediction was 0.06 sec in each case. Thus, the prime requirement of the algorithm is satisfied. As in the previous example, one needs at least one sample during the interval when the disturbance is active. Here also the probability of having no sample is during that interval is low.

C. Case Study 3

In this example a continuously-stirred tank heater (CSTH) given in [3] is simulated. Here the objective is to control the temperature of the fluid T is by manipulating the flow of liquid into jacket F_j around the heater. Like in previous case studies noise is added to the system in order to make simulation realistic. The time constant of the system is obtained as τ_p is 12 min. Hence the sampling time chosen to be $T_s = 2$ min and value of N is chosen as 20 samples. The time profile of the controlled variable resulting from the implementation of the proposed algorithm, for set-point changes, are shown in Figures 5. For better readability the plot of process variable with the locations of true measurements is shown in Figure 6. For comparison purpose, the profile of controlled variable for no missing data case is also shown in Figure 5. The controller tracks the setpoint well even in presence of high percentage of missing data. The time taken to compute each missing sample is 0.15 sec which is negligible when compared with sampling time.

The three case studies demonstrate the potential of the proposed method for control in presence of missing data, in both servo and regulatory cases.

D. A numerical study of stability of the proposed algorithm

As discussed at the end of §III stability of the algorithm is verified using Monte-Carlo simulations and is demonstrated on CSTH example described in previous case study. The experiment is repeated for 100 times and integral absolute error (IAE) value obtained after each run is plotted in Figure 7. The IAE is calculated using the formula

$$\text{IAE} = \sum_{k=0}^{q-1} |T - T_{sp}| \tag{14}$$

where T_{sp} is setpoint of the temperature and q is simulation time.

From the plot, it is evident that the IAE is almost constant in all simulations which proves that the algorithm is converging in all iterations, there by ensuring the stability of the algorithm.

V. CONCLUSION

In this paper a method was presented to estimate missing data online using CS techniques. In addition we shown the performance of the method with an application to classical PID control in presence of missing data. The proposed algorithm is demonstrated using three simulated examples. All the case

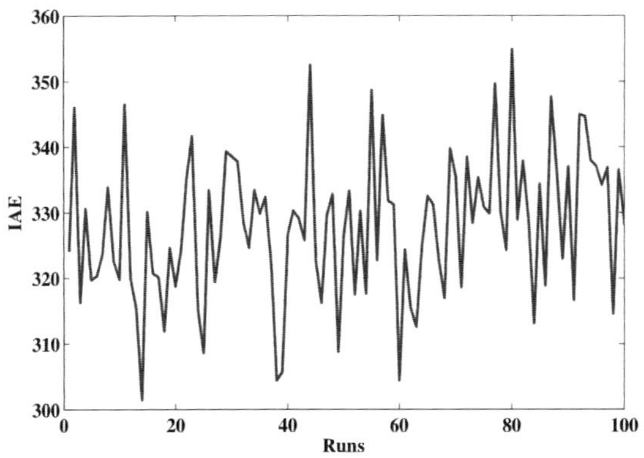

Fig. 7. Runs vs IAE

studies demonstrate the efficacy of the proposed method and shows comparable performance with full data case. At last we discussed the stability of the proposed algorithm and is proved using Monte-Carlo simulations on CSTH. Future directions include developing theoretical guarantees of the stability and extending the proposed algorithm for control of large scale systems.

REFERENCES

[1] T. Bohlin. On the problem of ambiguities in maximum likelihood identification. *Automatica*, 7(2):199–210, March 1971.

[2] Emmanuel J. Candès. The restricted isometry property and its implications for compressed sensing. *Comptes Rendus Mathematique*, 346(9–10):589 - 592, 2008.

[3] Denis Dochain. Process dynamics modeling, analysis, and simulation. *The Canadian Journal of Chemical Engineering*, 77(1), 1999.

[4] D.L. Donoho. Compressed sensing. *Information Theory, IEEE Transactions on*, 52(4):1289–1306, April 2006.

[5] M. Elad and M. Aharon. Image denoising via sparse and redundant representations over learned dictionaries. *Image Processing, IEEE Transactions on*, 15(12):3736–3745, 2006.

[6] Michael Elad. *Sparse and Redundant Representations: From Theory to Applications in Signal and Image Processing*. Springer Publishing Company, Incorporated, 1st edition, 2010.

[7] D. Guo, L. Zicheng, Q. Xiaobo, and L Huang. Sparsity-Based Online Missing Data Recovery Using Overcomplete Dictionary. *Sensors Journal, IEEE*, 12(7):2485–2495, 2012.

[8] Di Guo, Zicheng Liu, Xiaobo Qu, Lianfen Huang, Yan Yao, and Ming-Ting Sun. Sparsity-based online missing data recovery using overcomplete dictionary. *Sensors Journal, IEEE*, 12(7):2485–2495, July 2012.

[9] J. Haupt, W.U. Bajwa, G. Raz, and R. Nowak. Toeplitz compressed sensing matrices with applications to sparse channel estimation. *Information Theory, IEEE Transactions on*, 56(11):5862–5875, Nov 2010.

[10] N. Hurley and Scott Rickard. Comparing measures of sparsity. *Information Theory, IEEE Transactions on*, 55(10):4723–4741, Oct 2009.

[11] Lennart Ljung. *System Identification - A Theory for the User*. Prentice Hall International, New Jersey, USA, 1999.

[12] T Moon. The Expectation Maximization Algorithm. *IEEE Signal Processing Magazine*, pages 47–60, 1996.

[13] A. O'Dwyer. *Handbook of PI and PID Controller Tuning Rules*. Imperial College Press, 2009.

[14] Satheesh Perepu and Arun K. Tangirala. An adaptive basis estimation method for compressed sensing with applications to missing data reconstruction. In *DYCOPS 2013*, pages 190–196, Mumbai, India, 2013. IFAC.

[15] C. R Rao, T Helge, Shalabh, and H Christian. *Linear Models and Generalizations*. 1995.

[16] J. Romberg and M. Wakin. Compressed Sensing: A Tutorial. In *IEEE Statistical Signal Processing Workshop,*, 2007.

[17] J.A. Tropp. Just relax: convex programming methods for identifying sparse signals in noise. *Information Theory, IEEE Transactions on*, 52(3):1030–1051, March 2006.

[18] Jinghua Zhong. Pid controller tuning: A short tutorial. *Mechanical Engineering, Purdue University*, 2006.

[19] J. G. Ziegler and N. B. Nichols. Optimum settings for automatic controllers. *Transactions of ASME*, 64:759–768, 1942.

WATERFALL®
Stronger Than Firewalls

Evolving cyber security best practices for network perimeter protection

Michael Firstenberg, Director of Industrial Security
Waterfall Security Solutions

WATERFALL®

Cyber Perimeter - How Secure are Firewalls Really?

Attack Type	UGW	Fwall
1) Phishing / drive-by-download – victim pulls your attack through firewall		
2) Social engineering – steal a password / keystroke logger / shoulder surf		
3) Compromise domain controller – create ICS host or firewall account		
4) Attack exposed servers – SQL injection / DOS / buffer-overflowd		
5) Attack exposed clients – compromised web svrs/ file svrs / buf-overflows		
6) Session hijacking – MIM / steal HTTP cookies / command injection		
7) Piggy-back on VPN – split tunneling / malware propagation		
8) Firewall vulnerabilities – bugs / zero-days / default passwd/ design vulns		
9) Errors and omissions – bad fwall rules/configs / IT reaches through fwalls		
10) Forge an IP address – firewall rules are IP-based		

Attack Success Rate: Impossible | Routine | Easy

Photo: Red Tiger Security

Firewall have been with us for 30 years now. The good guys and the bad guys both know how to defeat them

 2

WATERFALL

Unidirectional Security Gateways: Server Replication

- Hardware-enforced unidirectional server replication
- Replica server contains all data and functionality of original
- External clients communicate only with replica historian
- 100% secure from online attacks from external networks
- Replicate historian servers, OPC servers, RDB servers, Modbus, etc.

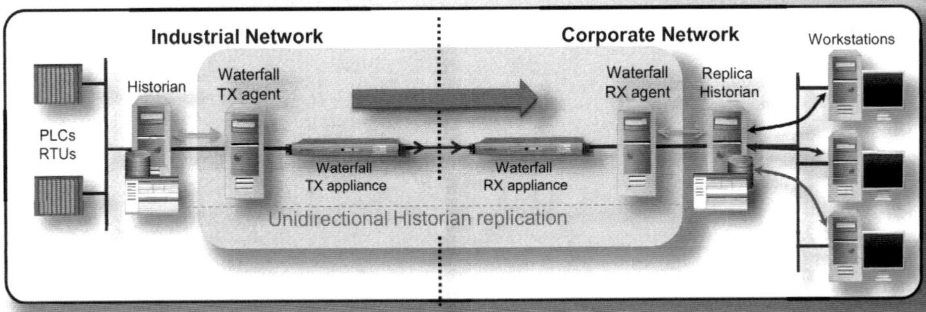

 3

WATERFALL

Modular Hardware Package

- Can use Waterfall host modules, or customer-supplied servers
- Supports dedicated TX/RX agent hosts, or shared, or virtual
- Supports all-in-one 1u rack-mount form factor, or each module in separate cabinet, or any combination
- Supports Windows and Linux
- User serviceable, user-expandable
- Standard: 2x dual power supplies
- Standard: 1Gbps

Modular, powerful, flexible, and secure gateway hardware system

 4

WATERFALL

Waterfall FLIP™ Defeats Interactive Remote Control

- Unidirectional Gateway whose direction can be reversed:
  - Chemicals / refining / mining / pharmaceuticals: batch instructions
  - Water systems: periodic security updates & anti-virus signatures
  - Remote unstaffed sites: substations, pumping stations
- Trigger: button / key, schedule
- Stronger than firewalls, stronger than removable media

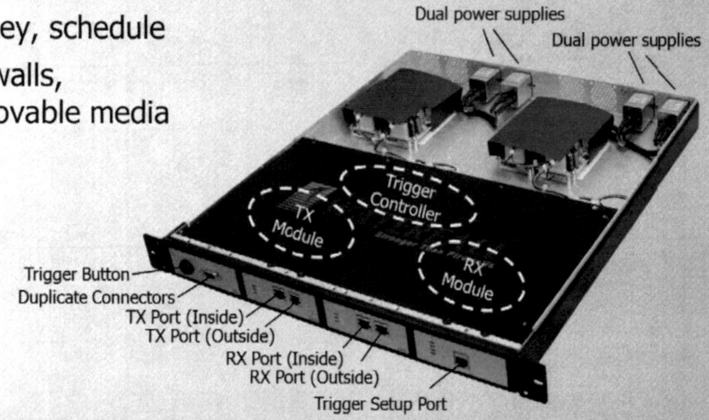

The FLIP is a Unidirectional Gateway that can "flip over"

 5

WATERFALL

FLIP: Stronger than Firewalls

- The FLIP is a Unidirectional Security Gateway – it can never be bi-directional
- The FLIP prevents interactive remote control – it can not FLIP fast enough to permit Remote Desktop or interactive SSH sessions
- Trigger mechanism cannot be subverted by network attacks
- Firewalls forward messages, the FLIP & Gateways do not
- No protocol-level attacks pass through – no fuzzing/buffer overflows. All communications sessions terminate in agent hosts.

FLIP: Stronger than firewalls

 6

WATERFALL

Application Data Control

- Powerful policy-driven application data control filters
  - Read and write value bounds checking, limiting
  - Limit commands, limit points & tags
- Apply policies to even proprietary, encrypted data streams
- Pass files & BLOBs to arbitrary third-party virus scanning, cryptographic signature verification & other tools

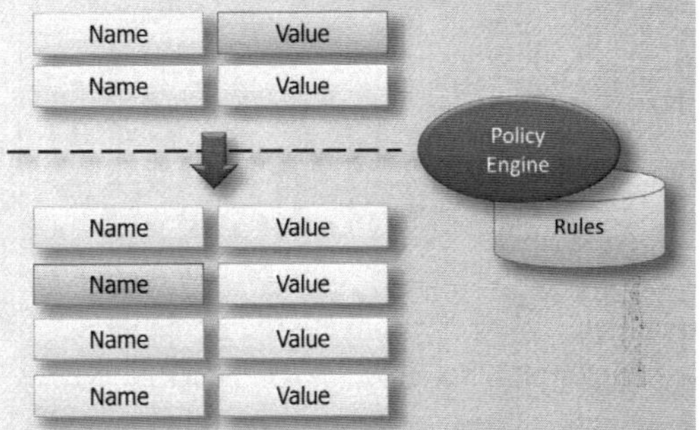

Waterfall ADC delivers on promises long made by next-gen firewalls

 7

WATERFALL

Evolving Best Practices – Unidirectional Gateways

NERC
NORTH AMERICAN ELECTRIC RELIABILITY CORPORATION

NERC CIP exempts unidirectionally-protected sites **from over 35% of requirements**

ICS-CERT
INDUSTRIAL CONTROL SYSTEMS CYBER EMERGENCY RESPONSE TEAM

DHS recommends unidirectional gateways in **security assessments** (ICS CERT)

NEI

NRC & NEI exempts unidirectionally-protected sites from **21 of 26 cyber-perimeter rules**

NIST
National Institute of Standards and Technology
U.S. Department of Commerce

NIST - unidirectional gateways prevent any connectivity of **traffic between domains** (800-82)

enisa

ENISA - unidirectional gateways provide **better protection than firewalls**

ISA IEC International Electrotechnical Commission

Unidirectional gateways – **limit the propagation of malicious code** (ISA SP-99-3-3 / IEC 62443-3-3)

 8

WATERFALL

Which of Our Networks are Expendible?

- Attacks only become more sophisticated over time
- Modern attacks routinely defeat firewalls & other security software
- Best practices are evolving – hardware-enforced Unidirectional Security Gateways are stronger than firewalls
- Absolute protection from network attacks originating on external neworks

FROST & SULLIVAN

"Waterfall's unique solutions have the potential to be the industry's next game changing standard"

So – which of our networks are expendable enough to protect with software alone?

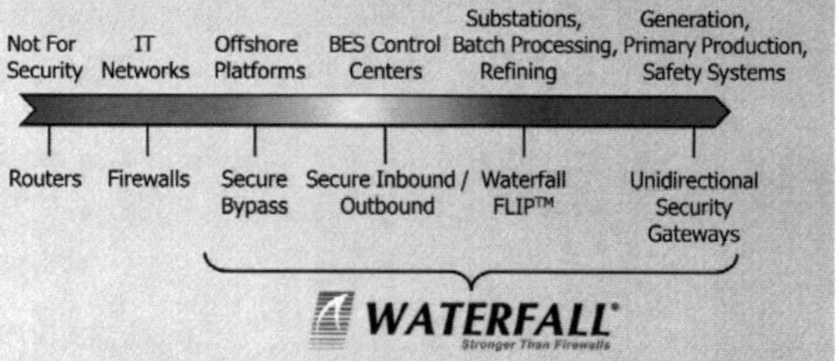

*** Formatting Issue - Best Available Paper/Graphic ***

398090 On Parabolic and Hyperbolic Diffusion Effects during Hybridization Kinetics with Convection in Microfluidic Chamber

Abstract:

The hybridization reaction between target molecules that is going to be studied with the probe molecules that are used for detection can be seen to be reversible reaction. Thus,

$$P + T \quad \overset{k_f}{\underset{k_r}{}} \quad P\text{-}T \tag{4.27}$$

where P is the probe molecules, T – target molecules and P-T the probe-target hybridized duplex that is formed on the substrate.

(4.28)

Where [P] is the concentration of the probe molecules, [T] is the concentration of the target molecules, [P-T] is the concentration of the probe-target duplex molecules, k_f is the rate constant of the forward reaction and k_r is the rate constant of the reverse reaction. The ratio of [P-T] to [P] provides a measure of the efficiency of the hybridization.

*** Formatting Issue - Best Available Paper/Graphic ***

0

$[P]_0$ is the concentration of the fresh probe molecules available for tagging in the beginning of the operation. This also can be seen to be the maximum concentration of probe molecules during the reaction process. When the target molecules are assumed to be in excess the rat expression in Eq. () can be seen to obey pseudo-first order kinetics. The rate would only vary with the concentration of the probe molecules and probe-target duplex molecules and the concentration of the target molecules can be lumped along with the rate constant, k_f and can be expected to remain constant during the reaction process. The rate expression can be simplified as;

(4.29)

Assuming that at t = 0, no probe-target duplex molecules were formed, i.e., $[P\text{-}T]_0 = 0$, Eq. () can be integrated as follows;

(4.29)

Where

Integrating Eq. ();

(4.30)

or

(4.31)

(4.32)

Where K is the equilibrium rate constant given by; . Eq. () is for the ideal case. This represents the highest rate possible for hybridization. Diffusion effects can result in a lower overall rate of hybridization.

The relative effects of intrinsic hybridization reaction rate and diffusion rate were evaluated using a Damkohler number. Damkohler number was defined by Pascault et. Al. [2008] as the ratio of maximal forword reaction rate to the maximum normal diffusion rate. When Da is greater than 1 they found the process to be diffusion limited.

Sharma [2005] has shown that when the dimensions of the chamber is small and the time of transport are low damped wave diffusion effects may become important. The relaxation times of the oligonucleotides may also have to be considered. The relaxation times can be expected to increase with the increase in size of the oligonucleotide. When the relaxation times are large and/or the microfluidic chamber size is small the concentration of the probe molecules can be expected to undergo subcritical damped oscillations. The probe molecules will have to meet the target molecules and form the duplex. The duplex molecules can be attached to the microarray substrate using linker molecules. Optimal probe concentration is discussed in Sharma [2009]. Quality of surface needed is also discussed. Single stranded DNA molecule will be 12 A^0 in diameter. So the P-T duplex molecules will be 24 A^0 in diameter. Due to major and minor grooves the effective diameter will be 20 A^0. The optimal target concentration is achieved when the spacing of target molecules are at least 20 A^0 each. Further packing will cause steric hindrance.

Consider the microfluidic chamber to be made of a finite slab with thickness of 2a (Figure 4.3). At one end of the surface the probe concentration is at the maximum and the constant wall concentration boundary condition can be assumed. The other end can be assumed to be impervious to the probe concentration. As the probe molecules diffuse through the target solution simultaneous reaction and damped wave diffusion effects may become important.

Figure 4.3 Microfluidic Chamber for P-T Duplex Formation

The governing equation for simultaneous damped wave diffusion, relaxation and reaction that obeys the irreversible first order kinetics, can be written as follows;

*** **Formatting Issue - Best Available Paper/Graphic** ***

(4.33)

Where C_p is the concentration of the probe molecules, t_r is the relaxation time of the probe molecules, D_{pt} is the binary diffusion coefficient of the probe molecules in target solution, k is the rate constant of the probe-target duplex formation assuming that the probe-target reaction is first order and irreversible and $k^* = kt_r$. The governing equation can be made dimensionless by the following substitutions;

(4.34)

The governing equation becomes;

(4.35)

The boundary conditions can be written as follows;

X = 0, u = 0 (4.36)

(4.37)

The governing equation needs to be made homogenous. The solution can be assumed to comprise of two parts, i.e., steady state part and a transient state part. Thus,

(4.38)

Pluggind Eq. (4.38) in Eq. (4,35) the governing equation can be written as two equations;

(4.39)

(4.40)

The governing equation to the steady state part is a second order differential equation with constant coefficients. The k* term will result in a integrating factor of another constant a_3. The solution to the second order differential equation with constant coefficients can be written as a sum of sinh and cosh hyperbolic functions. The integrating factor a_3 can be seen to be -1 by a $g = (1+ u^{ss})$ substitution. Thus;

(4.41)

Applying the boundary conditions to the solution a_2 can be seen to be 1 from the X = 0, u = 0 condition. Applying the impervious boundary condition at x = 2a;

(4.42)

(4.43)

(4.44)

The solution to the transient part of the dimensionless concentration can be obtained by the method of separation of variables. The dimensionless transient concentration can be expressed as a superposed product of decaying exponential and wave concentration. This can also be arrived at by multiplying Eq. () by e^{nt}. By choice of n and grouping the terms the wave concentration can be seen to be;

***** Formatting Issue - Best Available Paper/Graphic *****

(4.45)

For , the wave concentration can be seen to obey;

(4.46)

The hyperbolic second order partial differential equation for wave concentration can be solved for by the method of separation of variables. Let W = g(t)f(X). Then Eq. () becomes;

(4.47)

The space domain solution can be written as follows;

(4.48)

From the boundary condition at X = 0, c_2 can be seen to be 0. From the impervious boundary condition at x = 2a;

(4.49)

(4.50)

Or, , n = 0, 1, 2, 3, 4,……. (4.51)

The time domain solution can be seen to be;

(4.52)

It was shown by Sharma [2005] that final condition in time can be applied and solutions within the frameword of second law of thermodynamics can be obtained. When t ®∞, what happens to the wave concentration ?

(4.53)

The dimensionless concentration u becomes 0 and the exponential becomes infinity. Although the product of 0 and infinity is of the indetermintate form of the fourth kind [Piskunov, 1965] it can be assumed that W will be finite or zero at infinite time and will not be infinity. Applying this condition to the time domain solution it can be seen that c_3 is 0. The generation solution for the transient concentration can be written as sum of infinite modified Fourier series and;

(4.54)

The initial condition can be used in order to obtain the c_n by invoking the principle of orthogonolity. C_n can be seen to be . When the relaxation times are large the concentration of the probe can be expected to undergo subcritical damped oscillations. This is when;

(4.55)

This can happen with more probe molecular types especially when a is small. This is the case for the microfluidic chamber in Figure 4.3. This is about 100 μm. k* also contains a contribution from relaxation time, t_r. For systems with slow reaction and higher diffusivity Eq. () will yield a lower threshold value for the relaxation time. For systems with large relaxation times the concentration profile can be written as follows;

400957 Dynamic Simulation for Poly-Ethylene Process By PC-SAFT Model

Abstract:

Current applications for polyethylene plant simulation, especially for Operator Training Simulators (OTS), safety features analysis in dynamic, as well as ethylene compressor control mechanism, require both model accuracy and simulation performance. However this cannot be achieved without proper selection of the thermodynamic model. Perturbed-Chain Statistical Associating Fluid Theory (PC-SAFT) [1] has been well established through its accuracy in modeling monomer solubility into polymer solutions, and hence placed itself as an excellent candidate for model accuracy of polyethylene plant simulation.

Questions for PC-SAFT come from the performance and reliability of the density solutions on vapor/liquid phase at extreme conditions such as high temperature and high pressure [2]. Here, we present a state of the art implementation and modification of PC-SAFT through DYNSIM, a dynamic simulation tool by Schneider Electric that achieves both model accuracy and simulation performance, with which clients can apply to execute various tasks in both steady state and dynamic analysis over the entire plant.

Detailed information for the implementation and modification of PC-SAFT would be discussed in the scopes of

- Super-linear approach when previous rigorous solution had been converged;
- Pseudo-density extrapolation near and over super-critical region

Demonstration by DYNSIM simulation with PC-SAFT at key regions of polyethylene plant would be displayed, followed by the potential upstream and chemical plants' applications.

[1] J. Gross and G. Sadowski, "Perturbed-Chain SAFT: An Equation of State Based on a Perturbation Theory for Chain Molecules", Ind. Eng. Chem. Res. 2001, 40, 1244-1260

[2] Privat R., R. Gani and J-N. Jaubert, "Are safe results obtained when the PC-SAFT equation of state is applied to ordinary pure chemicals?", Fluid Phase Equilib. 2010, 1, 76-92.

399610 Advances in Liquid Level Systems for Process Control Education and Researches

Abstract:

Process control is closely related with static and dynamic characteristics of processes. Experiments will improve the process control course. For this, small experimental systems without producing wastes have been proposed in AIChE spring meeting 2012 (Apr. 1-5, Houston, Texas, U.S.A.) by the authors. Many advances have been made since then and will be presented here.

For the purpose of process control education, liquid level systems are adequate, but most of liquid level systems available are too simple to illustrate various process static and dynamic characteristics (Astrom, K.J. and Ostberg, A.B., "A Teaching Laboratory for Process Control," IEEE Control Systems Magazine, 6, 37-42, 1986). On the other hand, the four tank system suggested by Johansson (Johansson, K.H. "The Quadruple-Tank Process: A Multivariable Laboratory Process with an Adjustable Zero," IEEE Trans. Control Systems Technology, 8, 456-465, 2000) provides a system with unstable zero, which gives a nontrivial control problem, and is found to be very useful for process control education and researches. Several research papers have also been available (for example, Mercangoz, M. and Doyle III, F.J., "Distributed Model Predictive Control of a Four-Tank System," J. Process Control, 17, 297-308, 2007). The four tank system requires four tanks, four manual valves, two actuators and two level sensors. Our previous liquid level systems have been less complex and have more functions, realizing most of process characteristics which are in process control textbook (Seborg, D.E., Edgar, T.F. and Mellichamp, D.A., Process Dynamics and Control, 2nd ed. Wiley, N.Y., 2004).

Functions of the proposed liquid level system include (1) Linear Element with 1st Order, 2nd Order, and so on, (2) Integral Element, (3) Time Delay Element, (4) Inverse Response Element (Unstable Zero Element), (5) S-shape Nonlinear Element (Mimic pH Process), (6) Input Multiplicity Element, (7) Hysteresis Element (Mimic Ignition of Exothermic Reactor), and (8) Limit Cycle Instability Element. Functions newly added in this work are (9) Unstable Element, (10) Under-damped Element, and (11) Large Negative Zero Element.

WATER INJECTION PUMP'S WASTE HEAT RECOVERY SYSTEM RETROFIT TO ENHANCE ENERGY EFFICIENCY

Zeeshan Farooq, Mahmoud Bahy Nour Eldin and Omar Rajeh, P&CSD Saudi Aramco, Dhahran

Abstract

Water Injection Pumps (WIPs) are essential to oil production, it injects water back into the reservoir to increase pressure and thereby stimulate production. The high pressure, high flow WIPs are big consumer of energy and if they are driven by Gas Turbines (GTs) the exhaust gases gives a lot of opportunity for energy integration. In this paper, Water Injection Pumps (WIPs) driven by Gas Turbines (GTs) using sweet gas as fuel is investigated for best option for waste heat recovery. We demonstrated that the proper utilization of waste heat with the available technology can lead to efficient WIPs system design that results in energy saving and energy based GHG emission reductions. Several technologies available today can retrofit old systems and design new expansions effectively to increase the overall system efficiency.

Introduction

Improving waste heat recovery systems in industrial facilities can reduce the overall demand for energy consumption. Waste heat can be utilized to reduce power import or also can even to produce power with the help of today's new technologies. In this paper we have system that consists of Water Injection Pumps (WIPs) driven by Gas Turbines (GTs) utilizing Sales Gas (sweet gas) as fuel. The GTs exhaust is utilized for Heat Recovery Steam Generators (HRSGs) to produce MP steam to be used as heating utility for Gas and Oil Separation Plants (GOSP). The GOSP has expansion plans to increase oil and gas production. The flue gas exhaust system is designed for 400^{0}F; this high exhaust temperature is used to enable use of back-up fuel gas (sour gas) in emergency cases when Sales Gas is not available. A new system is proposed to produce more steam by utilizing heat available in the flue gas, which can be switched back to the current system. In order to maximize the waste heat recovery efficiency the produced steam is used for Chilling in Gas Train, inlet cooling to GTs and power production via Condensing turbine. This case study demonstrate that huge savings are possible when the current design i.e. exhaust flue gas temperature of 400^{0}F shall be lower to exhaust temperature of 250^{0}F.

Process Description

The GOSP presented in this study consists of four 3-phase Separator and Gas Compression units and two Gas Trains consists of Dehydration (TEG) and NGL units. This facility also has planned incremental expansion to have one more GOSP and one more Gas Train. System can be represented pictorially as follows;

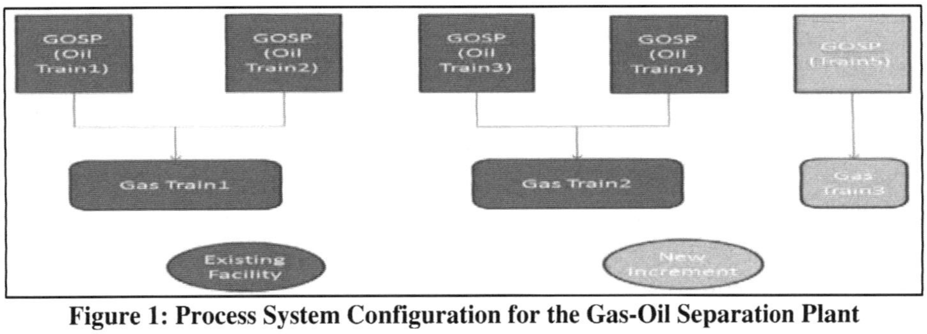

Figure 1: Process System Configuration for the Gas-Oil Separation Plant

The main process utilities are mention below and represented pictorially in Figure 2.

1. Power (Exported from outside)
2. MP Steam (Produced from HRSGs and Let down for LP steam)

3. Cooling Water (Cooling Tower)
4. Air (Fin-fans)
5. Chilling (Vapor Compression Refrigeration Cycle)

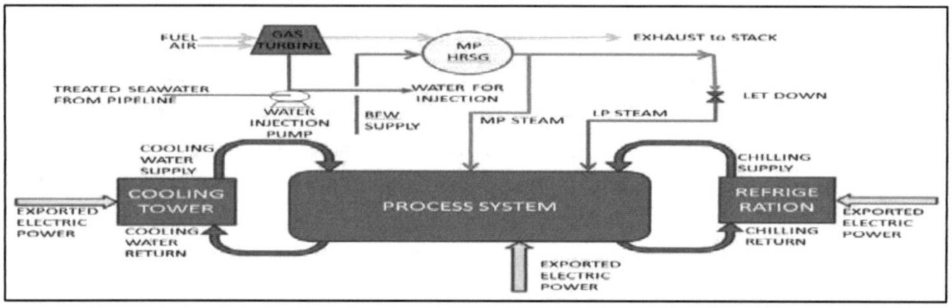

Figure 2: Utility System Configuration for the Gas-Oil Separation Plant

The system shows that the heat (steam) provided to the process is almost free (provided from HRSGs (Heat Recovery Steam Generators)) and any heat integration will results in saving steam which is the only heating utility results in no monetary savings. It is very clear from the system big picture that any steam saved won't result in appreciable monetary savings; hence no steam saving initiatives will justify investment. This lead us to the conclusion that any saved steam needs to be utilized in a better way to generate monetary savings to the facility. Moreover, it is noteworthy to pay attention that facility requirement for direct power as well as power for vapor compression cycle is large. Therefore, there is a need to explore the available technologies to utilize steam for power and chilling requirement.

Problem Statement

Given WIPs system with waste heat recovery in place and new expansion plans, identify options/scenarios for both existing and new waste heat recovery system to improve the overall system's efficiency.

Solution Approach

Solution approach adapted is to do targeting for the GT exhaust flue gases to produce HP, MP steam. Utilize excess steam (remaining after fulfilling process heating requirement) for all possible configurations (with available proven technologies) to make all possible cases to be evaluated with CHP model in terms of Net Present Value (NPV).

Water Injection Pumps (WIPs) in consideration are Gas Turbines (GTs) driven pumps. Steam generation is done through waste heat boilers, utilizing GTs exhaust to produce MP (560 mpph, 140 mpph per WIP) steam. Facility steam requirement is Medium Pressure (512 mpph) and Low Pressure (48 mpph) Steam, MP steam is let down to be utilize as LP steam through let down stations. CHP system is simple and no power can be produced with this arrangement. Present facility have 5 WIPs driven by GTs with four GTs exhaust is utilized in HRSGs to produce steam. Future facility as per planned expansion shall have seven WIPs all driven by GTs and all exhaust shall be utilized in HRSGs to produce MP steam with same arrangement results in inefficient system. Currently all steam produced from HRSGs is utilized by the facility but with the new increment there is unbalance in generation and consumption. The MP steam demand increases from 560 to 870 mpph and LP steam demand increases from 48 to 60 mpph.

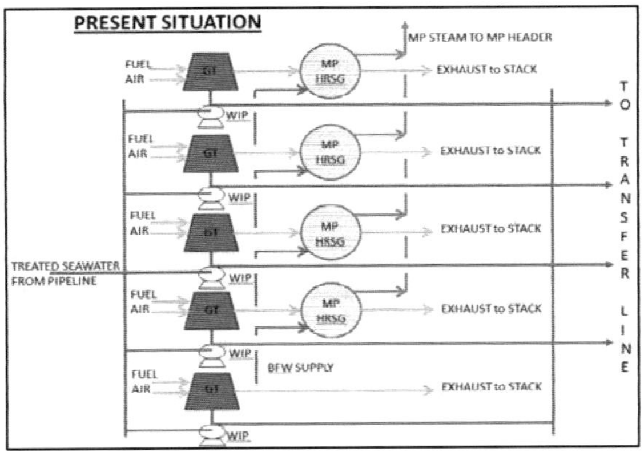

Figure 3: Present Steam Generation Arrangement

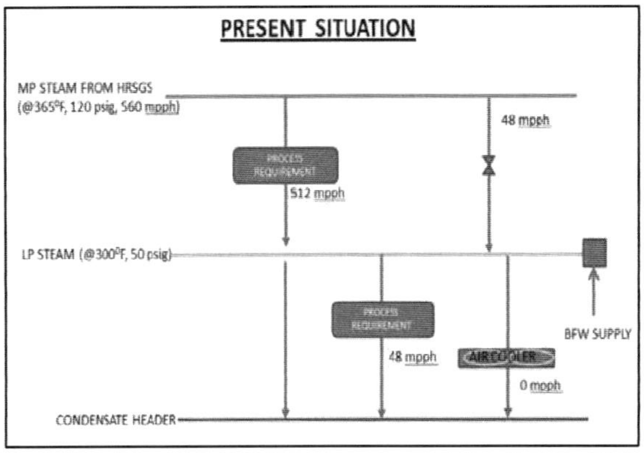

Figure 4: Present CHP Arrangement for the facility

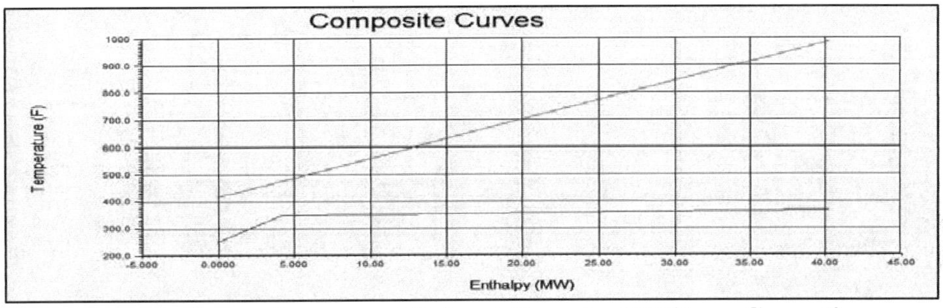

Figure 5: Composite curves of Exhaust Gases for MP Steam Generation

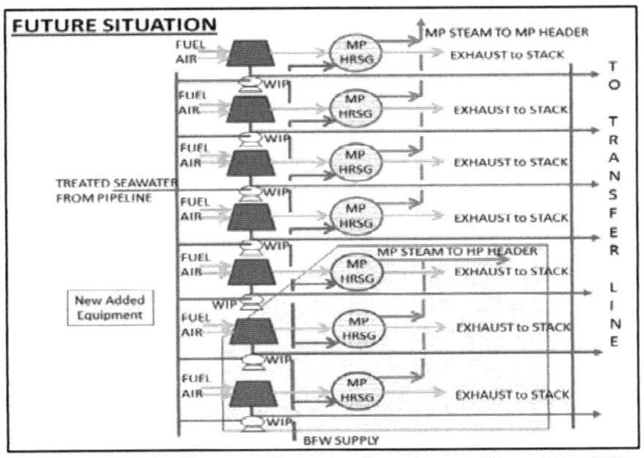

Figure 6: MP Steam Generation from New HRSGs

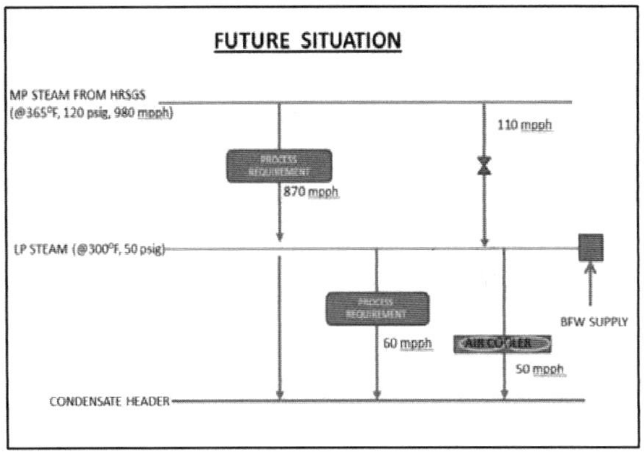

Figure 7: CHP for MP Steam from New HRSGs

If same arrangement is followed (i.e. same amount of MP steam is produced in the new HRSGs) in the future expansion which is considered as base case to compare all other alternatives schemes, not a perfect steam generation-consumption match is possible (figure 7). There is a need to optimize the system, as explained earlier the method used is to target the exhaust gases to produce steam and facility CHP model is utilize to evaluate all possible schemes. Targeting of exhaust gas (as shown in figure 5 & 8 Composite Curves for the system) shows that there is potential to produce HP steam of same amount (140 mpph per HRSGs) instead of MP steam which can be let down to produce power. With this information there are two cases which can be proposed as options to design the facility optimal steam system.

1. New HRSGs can produce HP steam instead of MP steam and
2. New HRSGs as well as existing HRSGs can be upgraded to produce HP steam

In the first option the new design HP HRSGs cost need to justify that additional investment for HP versus MP HRSGs while in second option cost of new HP HRSGs as well as cost of upgrading existing HRSGs will need to justify additional power production. These are the options which need to be optimized through evaluation of Capital Expenditure (Capex) and Operating Expenditure (Opex) to get results in terms of Net Present Value (NPV).

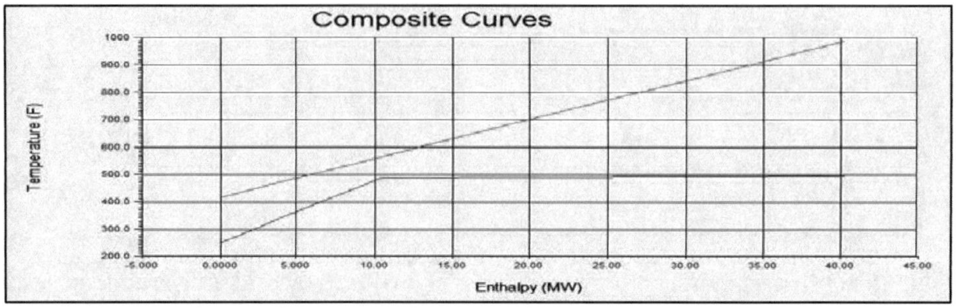

Figure 8: Composite curves of Exhaust Gases for HP Steam Generation

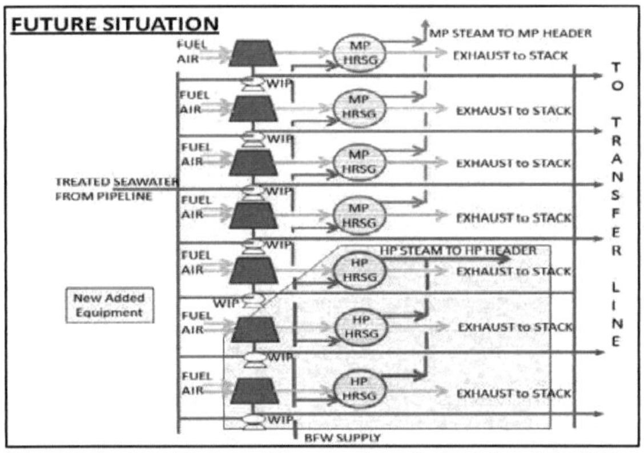

Figure 9: HP Steam Generation from New HRSGs

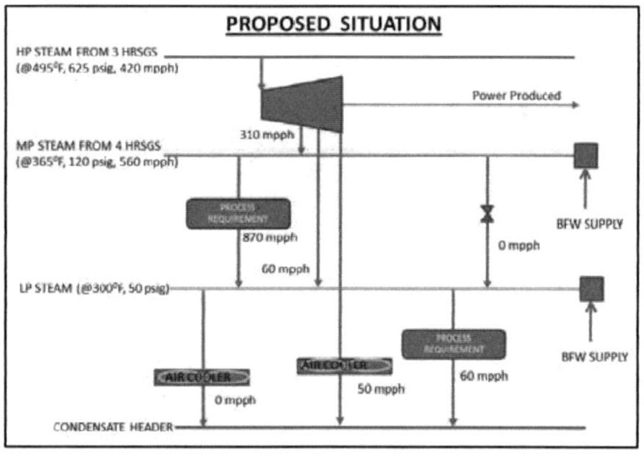

Figure 10: CHP for HP Steam from New HRSGs

Both options are quite similar with only difference, that in the first option we produce HP steam only with new HRSGs and have limited power production. Second option, with the upgrade of existing HRSGs we can produce more power but needs extra capital to upgrade the existing HRSGs to produce HP steam.

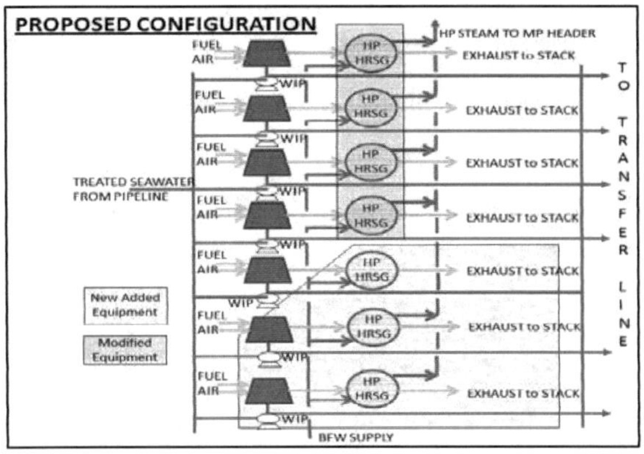

Figure 11: Max. HP Steam Generation Arrangement

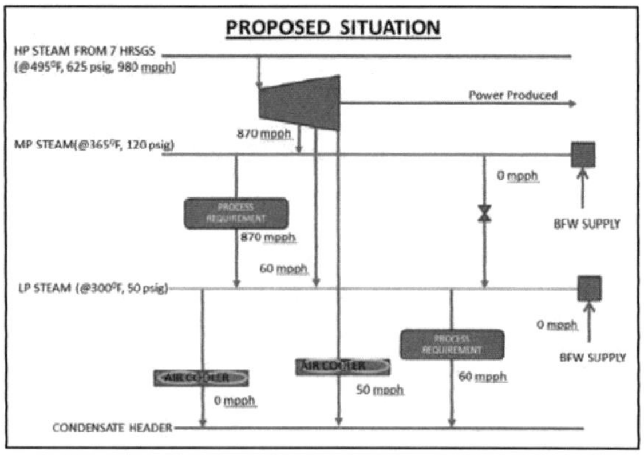

Figure 12: CHP for Max. HP Steam Generation

Of course there might be some options which can be proposed based on the utilization of the LP steam for some of the application requiring chilling but the quantity of excess steam is small to completely satisfy duty require by any chillers. Taking part load from the mechanical chillers will results in two systems which is very expensive especially for the same temperature range. Thus we can conclude that only these two options available to be evaluated in terms of NPV for the optimization.

Results and Discussion

As the exhaust temperature of the flue gas is very high (418°F) and needs to re-consider as we are using sweet gas as fuel having dew point below (250°F). The constraint of 400°F is due to utilization of sour gas in case of emergency when sales gas is not available, rarely occurring case (no incident till date from the plant start-up, more than 5 years). This lead to design a system utilizing heat wasted in flue gas and it was proposed to have a system which utilize the heat and switch back to exhaust flue gas at higher temperature when sour gas is in service. The proposed design in figure 14 can fulfil the requirement, though the proposed design is shown with blower but it could be designed without blower and depend on actual site arrangement and HRSGs in service etc. In this design flue gas leaving HRSGs shall be reduced to below 300°F if the fuel used is sweet gas but it would by-pass one of the HRSGs to be exhausted at above 400°F if the fuel gas used is sour gas. Targeting with new condition shows that we can produce additional 30 mpph MP steam per HRSGs (i.e. 170 mpph instead of 140 mpph) or we can produce 140 mpph per of HP steam and 30 mpph of MP steam from exhaust of each WIP.

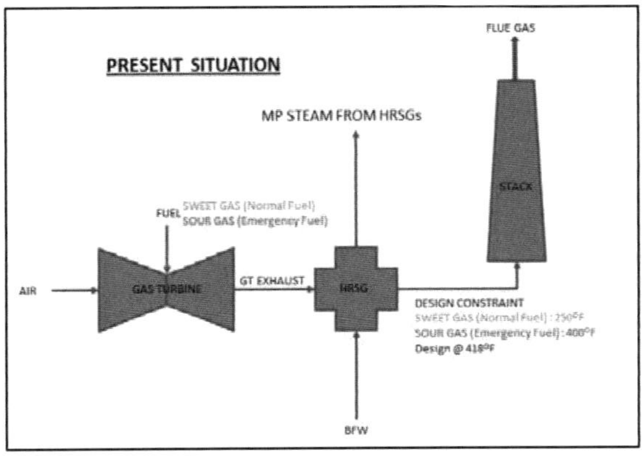

Figure 13: Present Steam Generation Scheme

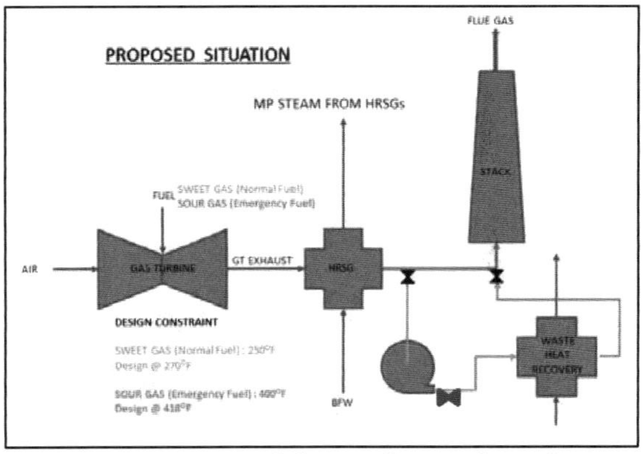

Figure 14: Proposed Steam Generation scheme

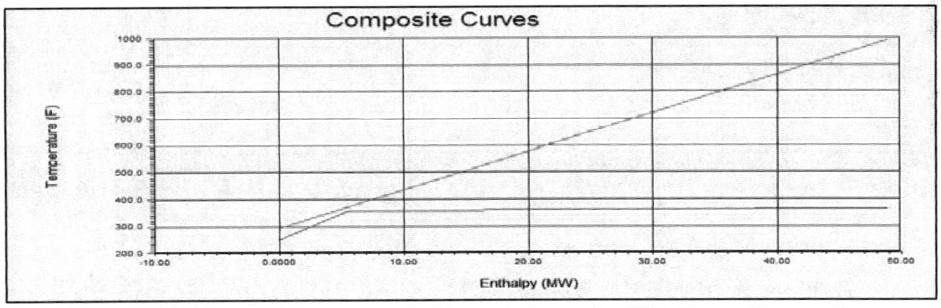

Figure 15: Composite curves of Exhaust Gases for MP Steam Generation

Based on the targeting results, there are possibly many cases which need to be evaluated for selecting optimum case, based on steam production following may be some additional options (in addition to earlier cases) for selecting optimum.

1. No change in existing HRSGs and new HRSGs produce more MP (170 mpph per HRSGs) steam
2. No change in existing HRSGs and new HRSGs can produce HP steam (140 mpph) and MP steam (30 mpph)
3. Modified existing HRSGs as well as new HRSGs to produce MP (170 mpph per HRSGs) steam
4. Modified existing HRSGs as well as new HRSGs to produce HP steam (140 mpph) and MP steam (30 mpph).

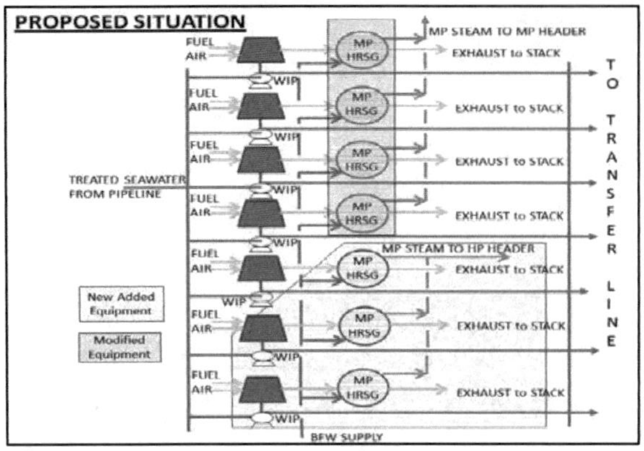

Figure 16: Max. MP Steam Generation Arrangement

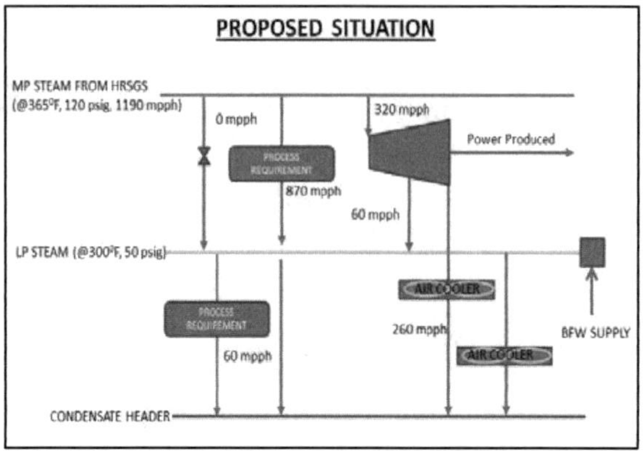

Figure 17: CHP for Max. MP Steam Generation

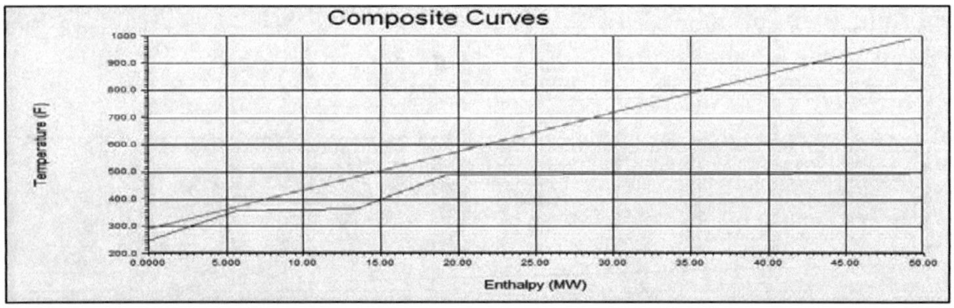

Figure 18: Composite curves of Exhaust Gases for HP & MP Steam Generation

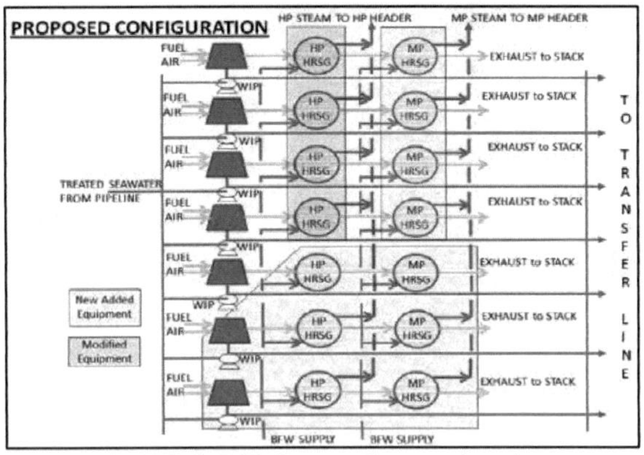

Figure 19: HP & MP Steam Generation Arrangement

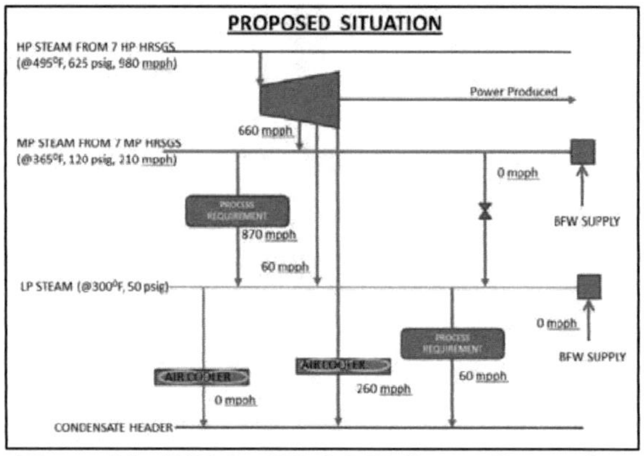

Figure 20: CHP for HP & MP Steam Generation

Now considering these cases clearly shows that power production from various options can vary from 0 (base case) to 14 MW (case shown in figure 19), which means different opex for each case. It also shows that the lowest investment cost is for the base case (i.e. putting same HRSGs design as the existing HRSGs) and highest capex for case shown in figure 19. These means all cases have potentials to be winning case depending on the price of the power produced versus capital required to produce it.

All the presented cases are considered with the assumption that the available heat shall be utilized to produce power (site power demand) and also with the assumptions that steam turbine is the best technology to do it. It is instructive to mention that even now optimization involving boilers to produce steam to use it as driver for one WIP can be explored. But utility system should complement process system and vice-versa, demand analysis results shows that along power requirement there is lot of refrigeration requirement for the site.

Site Wide Impact

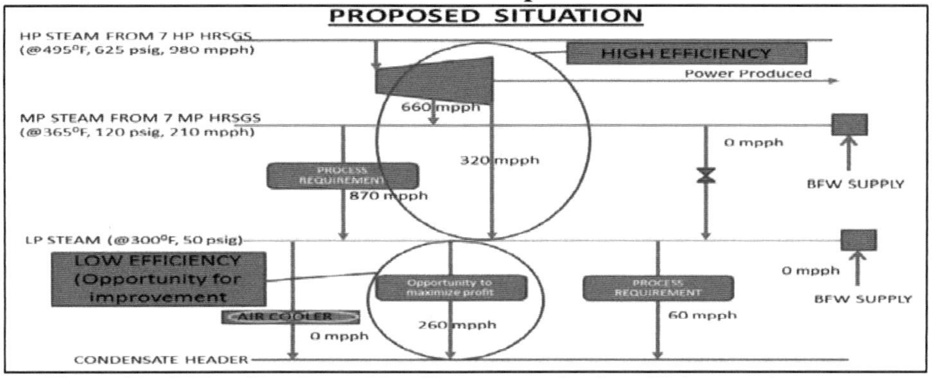

Figure 21: CHP utilization to identify opportunity for efficiency enhancement

Now considering site-wide impact for the proposed design, it is evident from the figure 21 that high grade heat utilization efficiency is quite high (back pressure turbine) can be further increase marginally. Like utilizing the available heat to produce steam and instead of utilizing steam to produce power, use it as steam driver for rotating equipments in the plant. It would be very interesting problem if the steam produced is enough to drive at least one WIP as this is not the case and there is not much potential to improve in utilization of high grade energy. Still there are lots of opportunities in the utilization of low grade energy to produce power via available technology like Organic Rankine Cycle (ORC), Kalina Cycle etc. As mention in Table 1 the amount of power produced from different technology is not very different for different technologies, which lead to the conclusion that initial investment or capex is the deciding factor for their utilization. Analysis of utility demand from the processes clearly indicates that there is lot of refrigeration requirement from the process which could be fulfilled by the available low grade energy utilizing Absorption Chillers or other available technologies to maximize the savings for the site.

Technology	Power Output
Condensing Turbine	~11 MW
Organic Rankine Cycle (ORC)	~7 MW
Kalina Cycle	~7-8 MW

Table 1: Utilization of Low Grade Heat for Power Production

Absorption chillers are systems which utilize heat to produce chilling capacity for low temperature cooling. The available steam quantity with new design constraints is more than enough for satisfying all chilling load for gas trains and inlet cooling for GTs. The amount of steam left after fulfilling facility chilling requirement is used to produce power via condensing turbine. Calculations are provided to show that the amount of savings increases many fold from initial saving just by power production.

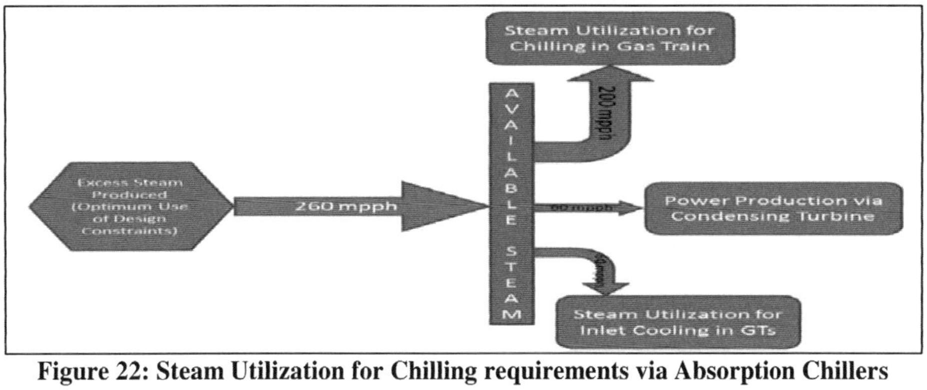

Figure 22: Steam Utilization for Chilling requirements via Absorption Chillers

Steam Utilization for Gas Train

Requirement by Trains & Stages	Operating Cooling		Design Cooling	
	MMBtu/hr	RT	MMBtu/hr	RT
First Stage Cooling for 1st Gas Train	55.73	4644	65.6	5467
First Stage Cooling for 2nd Gas Train	55.73	4644	65.6	5467
First Stage Cooling for 3rd Gas Train	27.87	2322	52.8	4400
Total First Stage Cooling	139.33	11610	184.0	15333
Second Stage Cooling for 1st Gas Train	23.53	1961	25.9	2158
Second Stage Cooling for 2nd Gas Train	23.53	1961	25.9	2158
Second Stage Cooling for 3rd Gas Train	11.77	980	21.0	1750
Total Second Stage Cooling	58.83	4902	72.80	6067
Total Cooling	**198**	**16513**	**257**	**21400**

Table 2: Cooling requirement for Gas Trains

Steam Utilization using absorption chiller	**Steam Consumed (lb/hr)**	**Refrigeration (RT)**
Refrigeration for 1st Stage for 2 Nos Existing Gas Train	77753	9288
Refrigeration for 2nd Stage for 2 Nos Existing Gas Train	32828	3922
Refrigeration for 1st Stage for New Gas Train	19438	2322
Refrigeration for 2nd Stage for New Gas Train	8207	980
Total Steam Utilized	**138226**	**16513**
Water Consumption	**m^3/hr**	**gpm**
Cooling Water Required (Refrigeration)	12192	53681
	m^3/hr	**gpm**
Make-up Water Required	**183**	**805**
Electrical Power Saving	**HP**	**kW**
E-51 (2fans @50 hp each) (New)	100	75
E-54A (6 fans with 40 hp each) (New)	240	179
E-54A (6 fans with 50 hp each) (New)	300	224
Refrigeration Compressor power for 2 Nos Existing Gas Train	25480	19000
Refrigeration Condenser power for 2 Nos Existing Gas Train	3840	2863
Refrigeration Compressor power for New Gas Train	9650	7196
Refrigeration Condenser power for New Gas Train	1440	1074
Total Power Saved	**41050**	**30611**
Power Consumption	**HP**	**kW**
Hot Water circulation pump (1pump @ 175hp) (New)	175	130
Cooling Water Tower Circ.Pumps (2 pump @ 750 hp) (New)	1500	1119
Cooling Water Circulation Pumps (2 pump @ 1500 hp) (New)	3000	2237
Cooling Water fans (20 fans @ 30 hp) (New)	600	447
Total Power consumption	**5275**	**3934**
	HP	**kW**
Net Power Saving in site Import	**35775**	**26677**
	HPH	**kWH**
Net Power Saving in site Import/year (@8760 operating hours)	**313389000**	**233694146**

Table 3: Calculation of potential Operating Cost Saving utilizing Absorption Chiller for Gas Trains

In table 2 gas train chilling requirement is shown and in table 3 the amount of power savings shown for case utilizing Absorption Chillers as compared to Vapor Compression system and the power saved is about 27 MW and mainly due to high power consumption in Propane Refrigeration System.

Steam Utilization for Inlet Cooling

Air Flow at ISO condition for 1 GT (MS5002D)	1113000	lb/hr
Energy Required to cool air for 1 GT from 104°F to 59°F	12.14	MMBTu/hr
Air Flow at ISO condition for 7 GT (MS5002D)	7791000	lb/hr
Energy Required to cool air for 7 GT from 104°F to 59°F	84.98	MMBTu/hr
Total Energy Required to cool all air from 104°F to 59°F	84.98	MMBTu/hr
Total Energy Required to cool all air from 104°F to 59°F	7082	RT
Steam Required for the cooling	59281	lb/hr
Water Consumption	**m^3/hr**	**gpm**
Cooling Water Required (Refrigeration)	5229	23022
Total Cooling Water Circulation Required	**5229**	**23022**
Make-up Water Required	**m^3/hr**	**gpm**
Make-up water for Cooling Tower	261	1151

Power Produced	HP	kW
Extra Power Generation (Inlet Air Cooling for 6 GTs and 1 Cogen)	72482	54050
Power Consumed	**HP**	**kW**
BFW pump (4pump @ 175hp each) (New)	700	522
Cooling Water Tower Circ.Pumps (2 pump @ 900 hp) (New)	182	136
Cooling Water Circulation Pumps (2 pump @ 1200 hp)(New)	2400	1790
Cooling Water fans (20 fans @ 25 hp) (New)	500	373
Flue Gas Blower (3 for three HRSGs @ 750 hp each)	2250	1678
Total Power Consumed	6032	4498
Fuel Consumption Increase (lb/sec)	5.48	
LHV for Fuel Gas (Btu/lb)	18310	
Fuel Consumption Increase (lb/year) (@8760 operating hours)	127837440	
Increase Energy Consumption (MMBtu/yr @8760 operating hours)	**2340704**	
Net Savings	**HP**	**kW**
Net Power Saving in site Import	**66450**	**49552**
	HPH	**kWH**
Net Power Saving in site Import/year (@8760 operating hours)	**430597563**	**321096560**

Table 4: Calculation of potential Operating Cost Saving utilizing Absorption Chiller for Inlet Cooling

Table 4 shows the effect of GTs inlet cooling which is saving of about 50 MW but with increase of 5.5 lb/sec increase in Fuel Gas. There are other technologies that can be utilize to produce similar results and might be options for optimization problem. This case study demonstrated that how changing a design constraint can lead to several options for optimization and consequently improving the efficiency of the overall system.

Results and Conclusions

In many industrial systems the design optimization is necessary. A lot of margins in adapted designs can lead to low overall system efficiency. Several beneficial process schemes may be unexplored if proper design limitations are not utilized by the designer. The system design limitation identification is very important to specify the proper process/utility system synthesis. In the WIPs case study presented in this paper, it is demonstrated that the system overall efficiency can be improved by designing system to exploit all the available energy in the flue gas. The extracted waste energy with the available Absorption Technology provided an opportunity to enhance the utilization of the available energy customized to the site demand. This enhanced overall system efficiency can only be attained via the proper consideration of the system limitations and designing system to those limitations.

References

Linnhoff, B., and Hindmarsh, E., "The pinch design method for heat exchanger networks" Chemical Engineering Science, Vol. 38, PP 745-763, (1983).

Linnhoff, B., "Pinch analysis-state of the art overview" Transactions Institution Chemical Engineering Research and Design, Vol. 71(A5), PP 503-522, (1993)

Biegler, L.T., Grossmann, I.E., Westerberg, A.W., "Systematic Methods of chemical process Design" Prentice Hall, New Jersey (1997)

Noureldin, M.B., "Methods for Heat Exchanger Network Energy Efficiency Assessment and Lifetime Retrofit", USP 8,150,560, (2012)

Noureldin, M.B., "Systems and Program Product for Heat Exchanger Network Energy Efficiency Assessment and Lifetime Retrofit", USP 8,150,559, (2012)

N.D.K. Asanta, X.X. Zhu, An automated approach for heat exchanger network retrofits featuring minimal topology modification, Computer & Chemical Engineering 20 (1996) 7-12

B.L. Yeap, D.I. Wilson, G.T. Polley, S.J. Pugh, Mitigation of crude oil refinery heat exchanger fouling through retrofits based on thermohydraulic fouling models, ChED 82A (2004) 53-71

399612 Control of Processes Whose Optimal Operating Conditions Are on the Saturation Edge

Abstract:

Hydrocarbon reforming can be used to produce economically hydrogen, a source for the proton exchange membrane fuel cell. The hydrogen-rich gas from hydrocarbon reforming reactors can have CO up to several thousand ppm level which poisons the anode catalyst of the fuel cell. The CO concentration need be removed below 10ppm. The preferential oxidation (PROX) is one of the simplest ways for this purpose. It uses the following oxidations:

$CO+0.5O_2$ --> CO_2

$H_2+0.5O_2$ --> H_2O

In above reactions, the CO oxidation should occur preferentially over the H_2 oxidation. The copper-cerium oxide catalyst of $CuO\text{-}CeO_2$ has been reported to have high activity and selectivity for CO oxidation. The CO PROX reactors with $CuO\text{-}CeO_2$ catalyst can remove CO from several thousand ppm to less than 10ppm. To design this PROX reactor, kinetics have been obtained by Lee and Kim ("Kinetics of CO and H_2 oxidation over $CuO\text{-}CeO_2$ catalyst in H_2 mixtures with CO_2 and H_2O," *Catal Today* 2008;132;109-116). Rigorous simulations have been done by Kim et al. ("Preferential CO oxidation over $CuO\text{-}CeO_2$ in excess hydrogen: Effective factors of catalyst particles and temperature window for CO removal," *Int. J. Hydrogen Energy* 2013;38;4429-4436). In this PROX reactor, limited O_2 should be added and O_2 should not exist in the exit stream for the optimal operations. A large amount of O_2 feed results in O_2 in the exit stream, reducing the efficiency of the PROX reactor. A small amount of O_2 feed results in high CO concentration in the exit stream and the objectives of the PROX reactor will not be met. The oxygen feed should be controlled on the edge just before O_2 appears in the exit stream.

There are several processes where optimal operational conditions are on the saturation edge as shown in the above PROX reactor. For example, the reactor temperature of polymerization reactor is given by the vaporization temperature of monomer and the optimal heat input is such that the reactor temperature should be on this saturation temperature edge. Additional heat increases the boil-up amount of monomer and the additional amount of cooling water should be needed. Some recycle systems with excess and limiting reagents give these control problems that maintain operating points on the saturation edges.

When the controlled variable is saturated, the variable has no information for feedback and the conventional control cannot be used. Control methods based on the sliding mode method and extremum seeking control method with coordinate system transformations are suggested and their performances are compared.

Presentation ID#: 399803

SYNTHESIS AND OPTIMIZATION OF AN INTEGRATED WATER AND MEMBRANE NETWORK SYSTEM WITH MULTIPLE ELECTRODIALYSIS REGENERATORS

Nielsen Y. Mafukidze, University of the Witwatersrand, Johannesburg, South Africa

Thokozani Majozi, University of the Witwatersrand, Johannesburg, South Africa*

Abstract

The shrinking supplies of freshwater globally, coupled with strict environmental regulations, have driven the manufacturing industry towards sustainable water management for the minimization of freshwater intake and wastewater generation. By using process integration and its enabling tools, this work considers the synthesis of an optimal water network with multiple regeneration capabilities. Development of the proposed framework is achieved by embedding a subnetwork of detailed electrodialysis models within a water network. Based on a superstructure and fixed-flowrate, the optimization problem is formulated as an MINLP model and solved in GAMS/DICOPT. To demonstrate the applicability of the proposed mathematical model a literature case study on a pulp and paper plant is presented and the results indicate a potential 12% savings in freshwater intake, 16% reduction in wastewater generated and a 14% saving in the total annualised cost for the entire network.

Keywords: Sustainable; Synthesis; Optimization; Electrodialysis

1. Introduction

Sustainability of human activities, primarily production and consumption, has been of growing concern in the process industry. Among the sustainability mega-forces, water scarcity has risen to the top of the corporate agenda for most entities in the industrial sector. In the face of this challenge, a number of industries are taking measures to become stewards of this vital resource through sustainable water management. With regards to the minimisation of freshwater intake and wastewater generation, process integration has presented enabling tools for sustainable water network design such as water recycle, reuse as well as regeneration-recycle and regeneration-reuse [1]. Optimization of water networks has become an essential instrument of sustainability as it allows for the realization of the full benefits, both economic and environmental, of using the aforementioned enabling tools.

The two optimisation approaches that remain irrefutable are the insight-based 'pinch' approach and mathematical programming. Most of the early work reported on water network optimisation was insight-based, which being graphically-based, focused on flowrate targeting and network design methods. Over the years, the technique has been refined and modified to extend its applicability to various systems found in the real world. Noteworthy is that these graphical methods offer low computation expense but are however limited to mass transfer based operations and grassroots design purposes [2]. Mathematical programming techniques involve the development of models based on mathematical relations that describe the system

---

* Corresponding author. Tel.: (+27) 11 717 7567/7384
Email address: Thokozani.Majozi@wits.ac.za

under analysis and the application of rigorous algorithms to obtain solutions. This technique became more prominent much later than pinch, as researchers realised its capability of treating rigorous, large-scale, complex systems, which pinch analysis is incapable of handling [2].

Within water networks, regeneration allows for the partial treatment of wastewater before being recycled or reused. An interesting observation is that for most of the work done so far on water-regeneration network optimization, the technologies of choice, number of regeneration units, arrangement of the regeneration train and regeneration design parameters are not set as decision variables; rather they are assumed to be known or oversimplified in a 'blackbox' fashion [3]. Contrary to that, Rangaiah and Wei [3] highlight that it is imperative to consider the technicalities of the system, including regeneration, such that complete structural and parametric (design) optimization of the given system is satisfied. Only a few researchers have considered the use detailed regenerator models for water networks. This includes El-Halwagi [4], Galan and Grossmann [5]. Khor *et al.*, [6] and more recently Yang *et al.* [7].

Electrodialysis (ED), the regeneration technology adapted in this work, has found large industrial scale use in the production of high quality industrial water as well as wastewater treatment of industrial effluents [8]. This water treatment process is based on the electro-migration and selective transport of ionic species induced by an electric current through a series of alternately arranged anion and cation-exchange membranes [8]. A majority of the insight on this particular process has been presented by Strathmann [8]. The current paper addresses the synthesis of a comprehensive water and membrane network that simultaneously gives an optimal design for the ED subnetwork. This is achieved by embedding a subnetwork of detailed ED models into a water network problem. Each regenerator model incorporates detailed cost correlations that are functions of the design variables.

2. Problem statement

In this work we consider a single-contaminant, integrated water network with multiple electrodialysis regenerators. The problem can be stated as follows. Given:

1. A set of fixed-flowrate water sources, *J*, with known flowrates and concentration of contaminant, which are flexible to recycle or reuse.
2. A freshwater source, *FW*, with variable flowrate and known contaminant concentration.
3. A set of fixed-flowrate water sinks, *I*, with known flowrates requirements and maximum allowable concentration of contaminants.
4. A wastewater sink, *WW*, with variable flowrate and maximum allowable concentration of contaminant. Effluent that requires further contaminant removal by chemical, biological or other means goes to this sink.
5. A set of regenerators, *R*, of different performances in terms of the removal ratio and the maximum allowable concentrations in to the regenerators. Flowrate into the regenerators is variable. Also given are some design parameters for ED units.

The objective is to synthesise an optimal integrated water network and to determine the design of the ED subnetwork that corresponds with the optimal water network. The optimal configuration is one that minimises freshwater use, generation of wastewater, regeneration energy and the capital and operating costs of both regeneration and the piping interconnections.

3. Optimization model

3.1. Superstructure representation

A superstructure approach that considers the water sources sinks and regeneration subnetwork simultaneously was used to formulate the water network problem considered in this work. The proposed superstructure includes all feasible interconnections between different units in the water-regeneration network and is presented in Figure 1. It includes a regeneration subnetwork as part of the water network. Provisions have been made for both parallel and series connection of regenerators as well as allowance for recycle of regenerator products back to the same regenerator feed. These provisions allow for a complete synthesis within the regeneration subnetwork and water network.

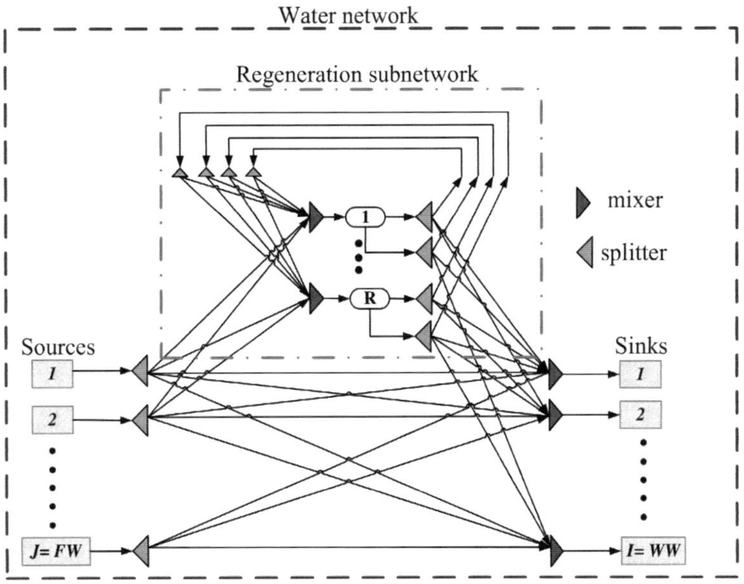

Figure 1: Superstructure representation of the water network.

3.2. Mechanistic electrodialysis regeneration network model

This section presents a mechanistic model formulation for a detailed design of an electrodialysis regeneration unit based on Tsiakis and Papageorgiou, [9]. The formulation includes mathematical equations for current required, voltage, membrane area, number of cell pairs, stack length, pumping energy and desalination energy. The regeneration network involves both capital and operational costs. The representative total annualised cost (*TAC*) function for the regeneration network is given by Equation (1). This function is included in the overall objective function of the water network, such that the energy consumption and subsequent cost of regeneration are minimised in conjunction with water consumption. To further improve flexibility, the model is enabled to only select the necessary regenerators that result in an optimal solution. This is achieved by prefixing the TAC function with a binary variable, y_r^{ED} which becomes zero when the respective unit is not activated.

$$TAC_r = y_r^{ED}\left[A_r\left(\frac{k^{mb}}{n}\right) + t^d k^{el} Q_r^{dil}\left(E_r^{spec} + E_r^{pump}\right)\right] \qquad \forall r \in R \qquad \text{[Eq. 1]}$$

3.3. Water Balances

To ensure connectivity between the regeneration network, the sources and the sinks, water balances are established based on the superstructure presented in Figure 1. These show how water from the different units split/mix into the other units. The maximum allowable load that each sink and regenerator tolerates is taken into account by means of contaminant balances. Logical constraints that ensure that streams from the same regenerator are not re-mixed in a single sink or recycled into the same regenerator are introduced. Additionally, to ensure that the maximum flowrates that can be accommodated by the piping are not exceeded, appropriate bounds are allocated to all the flowrates. These bounds are chosen based on practically feasible flowrates, such that any stream with a flowrate beyond these limits is not allowed to exist.

3.4. Objective function

The objective is to minimise the total annualised cost of the water network, which comprises the freshwater cost, wastewater treatment cost, annualised regeneration cost, as well as the capital and operating cost of piping. To account for capital and operational cost of piping, the length of pipe run, material of construction and linear velocities within the pipes were considered. The formulation of the objective function is expressed in Equation (2), where A^f is an annualisation factor.

$$obj = \min\left\{\begin{array}{l}\left(C^{FW}FW + C^{WW}WW\right)t^d + \sum_{r=1}^{R} TAC_r + \\ A^f\left[\begin{array}{l}\sum_{j=1}^{J}\sum_{i=1}^{I} D_{j,i}^s\left(\frac{pQ_{j,i}^s}{3600\cdot v} + q\cdot y_{j,i}^s\right) + \sum_{r=1}^{R}\sum_{i=1}^{I} D_{r,i}^d\left(\frac{pQ_{r,i}^c}{3600\cdot v} + q\cdot y_{r,i}^c\right) + \\ \sum_{r=1}^{R}\sum_{i=1}^{I} D_{r,i}^d\left(\frac{pQ_{r,i}^d}{3600\cdot v} + q\cdot y_{r,i}^d\right) + \sum_{r=1}^{R}\sum_{r'=1}^{R} D_{r,r'}^y\left(\frac{pQ_{r,r'}^x}{3600\cdot v} + q\cdot y_{r,r'}^x\right) + \\ \sum_{r=1}^{R}\sum_{r'=1}^{R} D_{r,r'}^y\left(\frac{pQ_{r,r'}^y}{3600\cdot v} + q\cdot y_{r,r'}^y\right) + \sum_{j=1}^{J}\sum_{r=1}^{R} D_{j,r}^k\left(\frac{pQ_{j,r}^k}{3600\cdot v} + q\cdot y_{j,r}^k\right)\end{array}\right]\end{array}\right\} \qquad \text{[Eq. 2]}$$

It is assumed that all the pipes share the same p and q parameter properties, stream velocity, v, and 1-norm Manhattan distance. The resulting mathematical model is a mixed integer nonlinear programming (MINLP) problem that was solved using GAMS/DICOPT.

4. Illustrative Example

The mathematical model developed is applied to a literature based pulp and paper plant case study [10]. This case study was chosen mainly because the industry is known to produce a lot of ionic effluent. This wastewater comes mainly from the bleaching section of the plant, where water is contaminated with chlorides and chlorates.

Table 1: Limiting data for the water sources and sinks.

Source, *j*			**Sinks, *i***		
j	**Flowrate (*t/h*)**	**Concentration (*mg/L*)**	*i*	**Flowrate (*t/h*)**	**Max. concentration (*mg/L*)**
1	890.1	0.003	1	1399.5	34.0
2	145.0	269.6	2	145.0	0.00
3	102.4	0.014	3	576.2	0.37
4	3099.0	499.8	4	3099.0	0.00
FW	∞	0	*WW*	∞	398.3

Table 1 shows the plant limiting data in terms of flowrate and contaminant (Cl) concentration for the sources and sinks amenable for wastewater integration. According to this table, the plant water network involves five water sources including the freshwater source and five sinks including the wastewater discharge.

To ensure a detailed analysis of the model, four scenarios are considered. Scenario 1 is water integration without regeneration. Scenario 2 is the case where water integration with regeneration involving two detailed ED regenerators is considered. However in this case, series connection and recycle within the regeneration subnetwork are forbidden just as in the case presented by Khor, *et al.*, [6]. Scenario 3 considers a water network with one detailed ED regenerator with capabilities of recycle within the regeneration subnetwork. Note that within this scenario two cases were considered, that is the case where the removal ratio of the regenerator is fixed at 0.733 and the case where it is variable. For the case where the removal ratio is fixed, an arbitrary value is chosen. Scenario 4 is similar to scenario 3 except that it has the capability of using two regenerators. Fixed removal ratios at 0.733 and 0.950 for the first and second regenerators respectively are chosen for the fixed removal ratio case.

Comparisons of the presented cases show that scenario 1 is the worst case possible. Scenario 4 proved to be the best and most preferable, since the largest amount of water was treated at the lowest total cost thereby minimizing the freshwater intake and wastewater generation.

Table **2** presents the optimum results for scenarios 4. It is evident from the results that by allowing the solver to choose the optimal configuration within the regeneration subnetwork in terms of recycles, series and parallel connections as well as making the removal ratio and number of regenerators required variable, there exist possibilities of obtaining better solutions rather than fixing them arbitrarily beforehand. In this way the designer gains more control of the unit performance by being able to stipulate the required membrane characteristics to the manufacturer. Additionally, since the technicalities of regeneration are included in the model, the performance of the regenerator, dictated by the removal ratio, directly influences the quantity of wastewater to be treated. As such allowing the optimiser to select both the performance and the quantity of water to be treated means a better optimum is guaranteed.

Table 2: Optimum results for water network for scenario 4.

	Fixed *RR*	**Variable *RR***
Freshwater flowrate (*t/s*)	1.006	0.989
Wastewater generation (*t/s*)	0.733	0.716
Total water regenerated (*t/s*)	0.699	0.586
Regeneration cost (million $)	0.562	0.353
Total cost (million $)	50.2	48.9
Optimal RR^*	-	r_1= 0.779 r_2= 0.770
CPU time (*s*)	6129	17230

The results also show that incorporating multiple regenerators can increase the chances of getting a better optimal solution, as long as the regenerators are not forced into the system. It is noteworthy that the proposed model is not limited to only two ED regenerators but can accommodate more as deemed necessary by the designer.

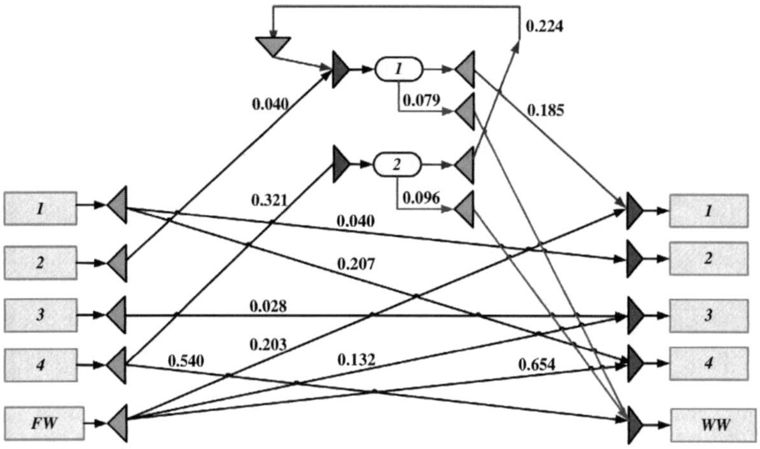

Figure 2: Optimum water network configuration for scenario 4 (variable RR).

Figure 2 shows the optimal network configuration and flowrates obtained for scenario 4. All flowrates are reported in t/s. According to Figure 2, that both series and parallel connection are vital for optimality. With this configuration in place, the plant is capable of generating savings of up to 12.7% in freshwater intake, reduction of up to 16.2% in wastewater generation and a 14.1% saving in the total annualised cost compared to the worst case scenario. The optimal regenerator design suggests an installation of two regenerators, r_1 and r_2, one with 2880 cell pairs and the other with 4356, amounting to a total membrane area of 3837 m^2 and 3921 m^2 respectively.

5. Conclusion

This work addresses the synthesis of a multi-regeneration water network by proposing an MINLP model that incorporates a detailed model of an electrodialysis regeneration subnetwork within a water network. To demonstrate the applicability of the proposed mathematical model, it is applied to a literature case study and solved using

GAMS/DICOPT. The results show that the use of a detailed model guarantees more accurate and reliable results in terms of water network synthesis as well as regenerator design. However, the nature and structure of the model proposed is quite complex, thereby resulting in very large computational times being incurred. Additionally, the system considered in this manuscript was limited to a single, ionic contaminant, whereas industrial effluent often contains multiple ionic contaminants as well as multiple non-ionic contaminants. As such, offering scope for future work, it would be desirable to extend the problem to allow for multiple contaminants, both ionic and non-ionic, using multiple regeneration technologies in order to be able to select the best available technology for the different streams of wastewater.

6. Nomenclature

Sets

J = $\{j|j = \text{water sources}\}$

I = $\{i|i = \text{water sinks}\}$

R = $\{r|r = \text{regenerators}\}$

Parameters

k^{el}	cost of electricity
k^{mb}	cost of membrane
C^{FW}	freshwater cost
C^{WW}	wastewater treatment cost
A'	annualisation factor
n	maximum equipment life
t^d	operating time per year
p	parameter for carbon steel piping
q	parameter for carbon steel piping
v	pipe linear velocity
$D^s_{j,i}$	Manhattan distance between source, j and sink i
$D^d_{r,i}$	Manhattan distance between regenerator, r and sink, i
$D^y_{r,r'}$	Manhattan distance between regenerators r and r'
$D^k_{j,r}$	Manhattan distance between source, j and regenerator, r

Continuous variables

TAC_r	total annualized cost for regenerator, r.
A_r	membrane area required by regenerator r.
E_r^{spec}	specific desalination energy required by regenerator r.
E_r^{pump}	specific pumping energy required by regenerator r.
Q_r^{dil}	Final diluate stream flowrate of regenerator, r
$Q^d_{r,i}$	Diluate flowrate to sink, i
$Q^c_{r,i}$	Concentrate flowrate to sink, i
$Q^k_{j,r}$	Flowrate from source, j to regenerator, r
$Q^s_{j,i}$	Flowrate from source, j to sink, i
$Q^x_{r,r'}$	Recycle from diluate stream to regenerator feed
$Q^y_{r,r'}$	Recycle from concentrate stream to regenerator feed
FW	Freshwater flowrate
WW	Wastewater flowrate

Binary variable

$y_r^{ED} = \begin{cases} 1 \leftarrow \text{if regenerator } r \text{ exists} \\ 0 \leftarrow \text{otherwise} \end{cases}$

$y^d_{r,i} = \begin{cases} 1 \leftarrow \text{piping exists between regenerator diluate and sink, } i \\ 0 \leftarrow \text{otherwise} \end{cases}$

$y^c_{r,i} = \begin{cases} 1 \leftarrow \text{piping exists between regenerator concentrate and sink, } i \\ 0 \leftarrow \text{otherwise} \end{cases}$

$y^x_{r,r'} = \begin{cases} 1 \leftarrow \text{recycle exists from regenerator diluate to regenerator, } r \\ 0 \leftarrow \text{otherwise} \end{cases}$

$y^y_{r,r'} = \begin{cases} 1 \leftarrow \text{recycle exists from regenerator concentrate to regenerator } r \\ 0 \leftarrow \text{otherwise} \end{cases}$

$y^k_{j,r} = \begin{cases} 1 \leftarrow \text{piping exists between source, } j \text{ and regenerator, } r \\ 0 \leftarrow \text{otherwise} \end{cases}$

$y^s_{j,i} = \begin{cases} 1 \leftarrow \text{piping exists between source } j \text{ and sink, } i \end{cases}$

0←otherwise

7. References

[1] M. M. El-Halwagi, Sustainable Design Through Process Integration: Introduction to Sustainability, Sustainable Design, and Process Intergration, Waltham, MA: Elsevier Inc., 2012.

[2] R. R. Tan, D. K. S. Ng, D. C. Y. Foo and K. B. Aviso, "A superstructure model for the synthesis of single-contaminant water networks with partitioning regenerators," *Process Safety and Environmental Protection,* vol. 87, pp. 197-205, 2009.

[3] G. P. Rangaiah and T. K. Wei, Eds., Advances in Process Systems Engineering: STOCHASTIC GLOBAL OPTIMIZATION, Techniques and Applications in Chemical Engineering, vol. 2, Toh Tuck, Singapore: World Scientific Publishing Co. Pte. Ltd., 2010.

[4] M. El-Halwagi, "Synthesis of Reverse-Osmosi Networks for Waste Reduction," *AIChE Journal,* vol. 38, no. 8, pp. 1185-1198, 1992.

[5] B. Galan and I. E. Grossmann, "Optimization Strategies for the Design and Synthesi of Distributed Wastewater," *Computers and Chemical Engineering,* pp. S161-S164, 1999.

[6] C. S. Khor, D. C. Foo, M. M. El-Halwagi and Tan R. R., Shah N., "A Superstructure Optimization Approach for Membrane Separation-Based Water Regeneration Network Synthesis with Detailed Nonlinear Mechanistic Reverse Osmosi Model," *American Chemical Society,* no. 50, pp. 13444-13456, 2011.

[7] L. Yang, R. Salcedo-Diaz and I. Grossmann, "Water Network Optimization with Wastewater Regeneration Models," *Industrial & Engineering Chemistry Research,* vol. 53, pp. 17680-17695, 22 September 2014.

[8] H. Strathmann, "Electrodialysis, a mature technology with a multitude of new applications," *Desalination,* vol. 264, pp. 268-288, 2010.

[9] P. Tsiakis and L. G. Papageorgiou, "Optimal design of an electrodialysis brackish water desalination plant," *Desalination,* vol. 173, pp. 173-186, 1995.

[10] I. M. L. Chew, Tan R., Ng D. K. S, Foo D. C., Majozi T. and Gouws J., "Synthesis of Direct and Indirect Interplant Water Network," *American Chemical Society,* vol. 47, pp. 9485-9496, 2008.

AN EFFECTIVE SYNTHESIS FRAMEWORK FOR INTEGRATED WATER AND MEMBRANE NETWORKS

Esther Buabeng-Baidoo, University of the Witswatersrand, Johannesburg, South Africa

*Thokozani Majozi**, *University of the Witswatersrand, Johannesburg, South Africa*

Abstract

Strict environmental regulations and social pressures have created the need for water and energy minimization in the process industries. This work therefore looks at the incorporation of a detailed reverse osmosis network (RON) superstructure within a water network superstructure in order to simultaneously minimize water, energy, operation and capital costs. The water network consists of water sources, water sinks and RO units for the partial treatment of the contaminated water. An overall mixed-integer nonlinear programming (MINLP) framework is developed that simultaneously evaluates both water recycle/reuse and regeneration reuse/recycle opportunities. The solution obtained from optimization provides the optimal connections between various units in the network arrangement, size and types of RO units, booster pumps as well as energy recovery turbines. The paper looks at four cases in order to highlight the importance of including a detailed regeneration network within the water network instead of the traditional 'black box' model. The importance of using a variable removal ratio in the model is also highlighted by applying the work to a literature case study which leads to a 28 % reduction in freshwater consumption and 80 % reduction in wastewater generation.

Keywords: Sustainable, Superstructure, Optimization, Synthesis, Reverse osmosis

1. Introduction

The scarcity of water and strict environmental regulations have made sustainable engineering a prime concern in the process and manufacturing industries [1]. Water minimization involves the reduction of freshwater use and effluent discharge in chemical plants. This is achieved through water reuse, water recycle and water regeneration. Water reuse involves the use of wastewater in other operations except the process where it was originally used. Water recycle, however, allows the effluent to be used in any process including the process in which it was produced. In water regeneration, the effluent is partially treated before it is recycled or reused in other processes. Partial treatment can be achieved by using water purification units often classified as membrane and non-membrane processes, e.g. RO membranes and steam stripping respectively [2].

The purification of water through membrane systems is an energy intensive process. The minimization of energy within the water networks is also needed for sustainable development. Energy usage within the water network is largely associated with the regeneration units (membrane units). In most published work, however, membrane systems

* Corresponding author: Tel: (+27) 11 7177567/7384
E-mail address: thokozani.majozi@wits.ac.za

have been represented using the "black box" approach [3-5] which uses a simplified linear model to represent the membrane systems [5]. This approach does not give an accurate representation of the energy consumption and associated costs of the membrane systems. A more rigorous representation of the regeneration unit is therefore needed [6].

The regenerator unit considered in this paper is the reverse osmosis membrane (RO). RO membranes are highly favored amongst other separation units due to their low energy consumption, ease of operation, high product recovery and quality [7]· RO membranes separate a water stream into a lean stream of low contaminant concentration known as permeate and a highly contaminated stream known as the retentate stream.

The rigorous design of RO units has been extensively studied in literature. Following on from Evangelista [8], El-Halwagi [7]introduced the 'state space approach' for the design of RO networks (RON). A superstructure representation of the RON was introduced with the objective of synthesizing a network with the optimal number of RO units, booster pumps and energy recovery turbines that resulted in minimal freshwater consumption.

There are two major approaches adopted in addressing water network synthesis namely, insight based techniques and mathematical model-based optimization methods. Insight-based techniques involve the water pinch analysis, which is a graphical method based on the concept of a limiting water profile which is the most contaminated water that can be fed into a particular operation. This method was first proposed by Wang and Smith [12]. Kuo and Smith [13] then expanded on the theory proposing a methodology for the design of treatment systems that consider both single and multiple contaminants. The computation burden of this method is, however, lower than that experienced by mathematical model-based optimization methods.

Mathematical optimization approach employs a superstructure which identifies an optimal configuration for the process from a number of alternatives. This idea was first proposed in the work by Takama *et al.* [16]. They proposed a nonlinear model that incorporates both water using and wastewater treating units for multiple contaminant systems. A significant amount of research has been conducted since then, exploring different techniques for modelling regenerators and developing solution strategies for the complex MINLPs [3, 5]. This method is however computationally expensive.

Khor *et al.* [2] presented a detailed membrane regenerator model which was incorporated into a water network superstructure (WNS). The work of Khor *et al*· [2] assumed a single regenerator with a fixed design, which implies that number of regenerators needed, number of pumps, number of energy recovery turbines were specified a priori.

The current work proposes a superstructure optimization for the synthesis of a detailed RON within a WNS. The overall mathematical formulation is used to minimize both water and energy simultaneously. A rigorous nonlinear RON superstructure model, which is based on the state space approach by El-Halwagi [7] is included in the WNS to determine the optimum number of RO units, pumps, turbines required for an optimal WNS. A fixed flowrate model that considers the concept of sources and sinks is adopted. The model takes into account streams with multiple contaminants. The idea of using variable removal ratio to describe the performance of the regenerators is also explored in this work.

2. Problem Statement

The problem addressed in this work can be stated as follows:

Given:

(i) A set of water sources, I, with known flowrates and known contaminant concentration. A set of water sinks, J, with known flowrates and known maximum allowable contaminant concentration.

(ii) A network of RO regenerators, Q, with known liquid recovery and design parameters.

(iii) A freshwater source, FW, with known contaminant concentration and variable flowrate.

(iv) A wastewater sink, WW, with known maximum allowable contaminant concentration and variable flowrate.

Determine:

(i) The minimal freshwater intake, wastewater generation, the energy consumed in the RON and the TAC.

(ii) The optimal configuration of the water network.

(iii) The optimal number of RO units, pumps and energy recovery turbines.

(iv) The optimal operation and design conditions of the RON such as feed pressure, number of hollow fiber modules per regenerator, stream distributions, separation levels etc.

3. Mathematical Model

The overall MINLP model is based on the superstructure represented in Figure 1, which is adapted from the work of El-Halwagi [7] and Khor *et al.* [2] . Figure 1 shows the incorporation of the RON superstructure with the WNS.

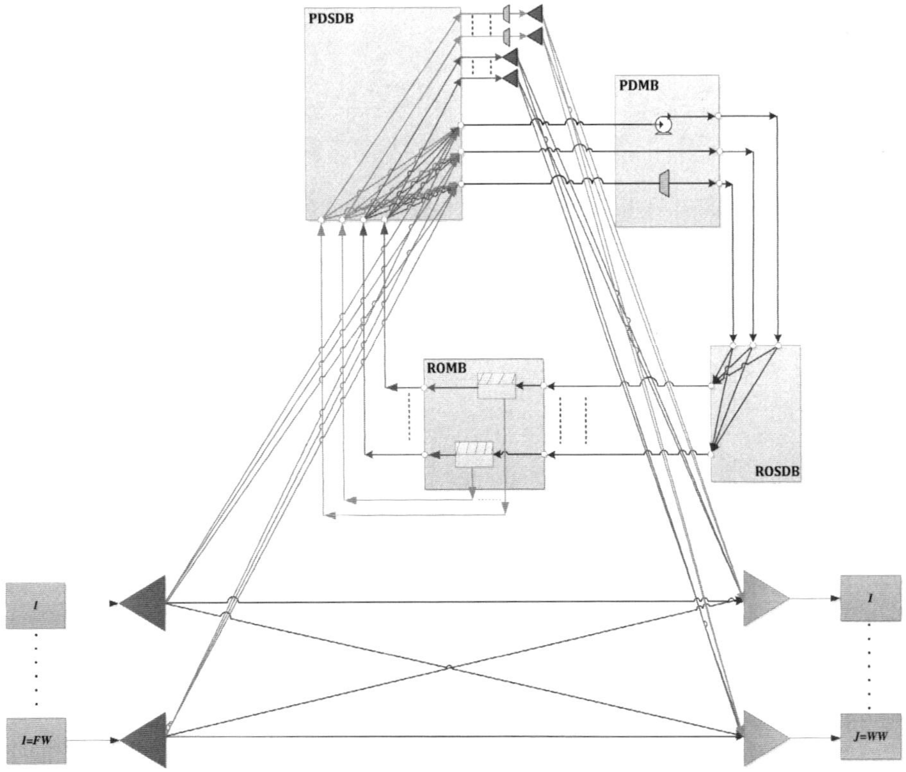

Figure 1. Superstructure representation of the RON superstructure within the WNS

3.1. *Water balances*

In order to ensure connectivity between the RON and the WNS, water balances are established based on the combined superstructure presented in Figure 1. From the diagram it can be seen that a water source can be fed to the RON, wastewater sink or to the water sinks. Figure 1 also shows that the water sinks receive water from the water sources, permeate and retenetate of the regeneration units as well as the freshwater source. In order to determine the existence of piping interconnections, logical constraints and discrete variables will be adopted.

3.2. *Objective function*

The objective function of the combined RON superstructure and WNS is used to minimize the overall cost of the regeneration network on an annualized basis which consists of:

(i). TAC of the RON

(ii). cost of freshwater (*FW)*

(iii). treatment cost of wastewater (*WW*)

(iv). capital and operation costs of the piping interconnection

The total annualized cost of the RON consists of the capital cost of RO modules, pump, and energy recovery turbines, operating cost of pumps and turbines as well as pretreatment of chemicals. The operating revenue of the energy recovery turbine is also considered in the determination of the TAC and is shown in constraint (1).

$$TAC(q,n) = +C^{pump}\left(\sum_{n=1}^{N}(P_n^i - P_n^a)F_n^a\right)^{0.65} + C^{tur}\left(\sum_{n=1}^{N}(P_n^i - P_n^o)F_n^a\right)^{0.43} \qquad \forall q \in Q \quad \text{[Eq.1]}$$

$$+C^{tur}\left(\sum_{j=1}^{J}(P_q^r - P_j^r)F_{q,j}^r\right)^{0.43} + C^{elec}AOT\left(\frac{\sum_{n=1}^{N}(P_n^i - P_n^a)F_n^a}{\eta_{pump}}\right)$$

$$-C^{elec}AOT\left(\sum_{j=1}^{J}(P_q^r - P_j^r)F_{q,j}^r\right)\eta_{turbine} - C^{elec}AOT\left(\sum_{j=1}^{J}(P_n^i - P_n^o)F_n^a\right)\eta_{turbine}$$

$$+C^{mod}\sum_{q=1}^{Q}N_q^s + C^{chem}AOT\sum_{i=1}^{I}F_{i,n}^d$$

Constraint (2) represents the objective function of the total regeneration network. It is also assumed that all the pipes share the same properties of *pc* and *qc* and a 1-norm distance D.

[Eq.2]

$$\min \begin{pmatrix} \sum_{q=1}^{Q}\sum_{n=1}^{N} TAC_{q,n} + AOTC^{water} FW + AOTC^{waste} WW \\ + AA\left(\sum_{i=1}^{I}\sum_{j=1}^{J} D_{i,j}\left(\frac{p_c F_{i,j}^{s}}{3600v} + q_c y_{i,j}\right)\right) \\ + AA\left(\sum_{q=1}^{q}\sum_{j=1}^{J} D_{q,j}^{p}\left(\frac{p_c F_{q,j}^{p}}{3600v} + q_c y_{q,j}^{p}\right)\right) \\ + AA\left(\sum_{q=1}^{q}\sum_{j=1}^{J} D_{q,j}^{r}\left(\frac{p_c F_{q,j}^{r}}{3600v} + q_c y_{q,j}^{r}\right)\right) \\ + AA\left(\sum_{i=1}^{I}\sum_{n=1}^{N} D_{i,n}^{d}\left(\frac{p_c F_{i,n}^{d}}{3600v} + q_c y_{i,n}^{d}\right)\right) \end{pmatrix}$$

The overall model results in a nonconvex MINLP due to the bilinear terms as well as the power function in the constraints.

4. Illustrative example

The above model was applied to a petroleum refinery case study based on the work presented by Khor *et al.* [2]. The model was implemented in GAMS 24.2 using the general purpose global optimization solver BARON which obtains a solution by using a branch-and-reduce algorithm. The network consists of 4 sources and 4 sinks. The limiting water data for the sources and sinks is given in Table 1.

Table 1. Limiting data for water network

Sources, *i*				**Sinks, *j***			
i	**Flowrate (kg/s)**	**Contaminant Concentration (kg/m³)**		*j*	**Flowrate (kg/s)**	**Max Contaminant Concentration (kg/m³)**	
		TSS	COD			TSS	COD
1	7.3	3.5	3.5	1	0.83	2.5	2.5
2	10.65	4	4	2	40	2	2
3	3.5	1	3	3	5.56	2.5	2.5
FW	—	2	1	*WW*	—	25	25

Table 2 shows the comparison between a case where multiple regenerators with fixed (case 1) and variable removal ratio (Case 2) where used. The removal ratio chosen by the model in case 2 was 0.97 for all contaminants instead the fixed value of 0.95 in case 1. Case 2 led to 3.12 % reduction in freshwater and 30.43 % reduction in wastewater generation in comparison with case 1. A 15.91 % reduction in the total network cost was also achieved. The large decrease in the total cost of the network in case 2 can be attributed to the high removal ratio which was selected by the model than the value that was initially predicted. The modeling of case 2 is however computationally expensive as can be seen in Table 2. Figure 2 shows the water network for case 2 with the corresponding flowrate for each stream. The best case used 15 HFRO modules per regenerator. The model selected two regenerators, two pumps and two energy recovery turbines as can be seen in Figure 2. It can also be seen that a parallel configuration of the network was chosen by the model. In comparison with the case where no regeneration was considered, case 2 leads to a 28 % reduction in freshwater consumption and 80 % reduction in wastewater generation.

Table 2. Summary of results for case 1 and 2

	Multiple regenerators	
	Fixed RR (Case 1)	Variable RR (Case 2)
Freshwater flowrate (kg/s)	28.87	27.68
Wastewater flowrate (kg/s)	3.91	2.72
Cost of regeneration (million $/year)	0.23	0.096
Total cost (million $/year)	1.32	1.11
Network configuration	Parallel	Parallel
Number of HFRO modules	37 for each regenerator	15 for each regenerator
CPU time (h)	6	54

The high computational time for solving the model in case 2 was due to the complexity of the problem as well as the large number of 0-1 variables. The model solves quicker when tighter bounds are imposed on the feed and retentate pressure. The use of the energy recovery turbines in the RON led to a reduction in the regeneration cost of the network, and as a result, a reduction in energy usage by the system was achieved.

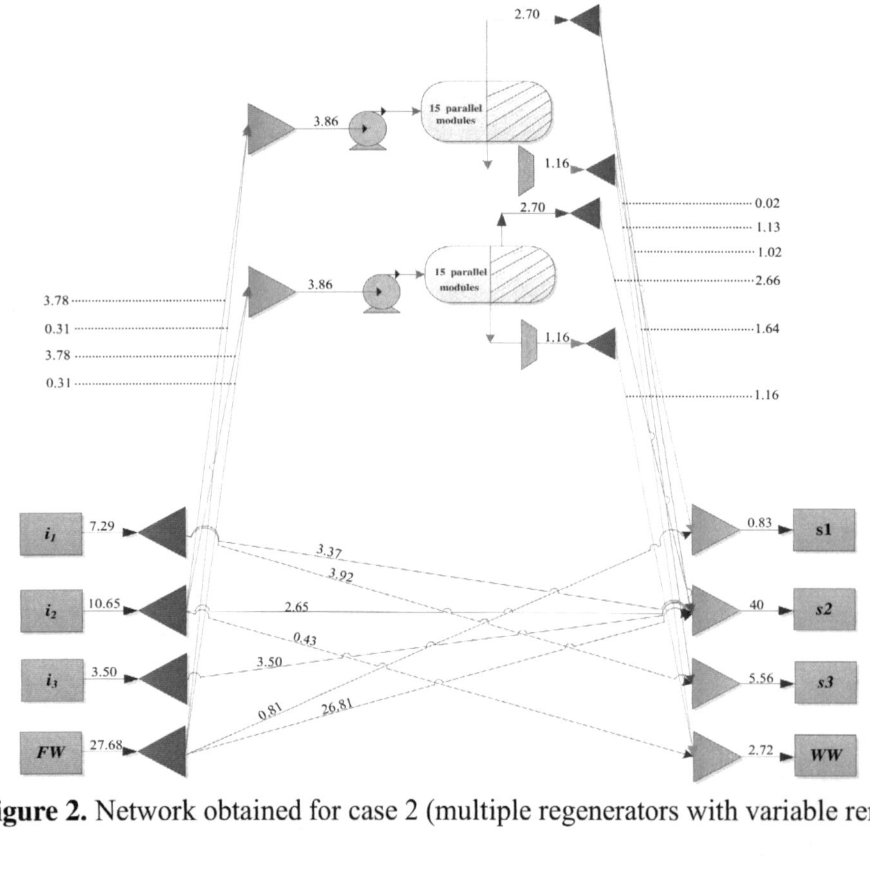

Figure 2. Network obtained for case 2 (multiple regenerators with variable removal ratio)

5. Conclusion

This paper has addressed the synthesis of a water regeneration network that incorporates the detailed synthesis of an RON. The proposed model was applied to a literature case study and was then solved using GAMS/BARON in order to highlight its practicality. The results show that the use of multiple regenerators in the water network, can lead to a reduction in the total cost of the network due to the significant reduction in freshwater consumption and wastewater generation. It can also be concluded that, there is a significant benefit in allowing the removal ratio in the model to be a variable as this has severe significant on the cost and structure of the network. Large computational times were however incurred due to the complex nature and structure of the model. It is also noteworthy that the proposed model was limited to one membrane technology. Multiple membrane technologies such as ultrafiltration can however be incorporated in the membrane network and thus offering a scope for future work.

6. Nomenclature

6.1. Sets

I={i|i=water source}

J={j|j=water sink}

M={m|m=contaminants}

Q={q|q=regeneration units}

6.2. Parameters

AOT	annual operating time
p_c	parameter for carbon steel piping based on CEPCI value of 318.3
q_c	parameter for carbon steel piping based on CEPCI value of 318.
v	velocity
η_{pump}	pump efficiency
$\eta_{turbine}$	turbine efficiency
$D^a_{i,j}$	manhattan distance between water source i and sink j
$D^p_{q,j}$	manhattan distance between regenerator q and sink j
$D^r_{q,j}$	manhattan distance between regenerator q and sink j
$D^d_{i,n}$	manhattan distance between source i and node n
C^{chem}	cost parameter for chemicals
C^{elec}	cost of electricity
C^{mod}	cost per module of HFRO membrane
C^{pump}	cost coefficient for pump
C^{tur}	cost coefficient for turbine
P^r_j	pressure of the retentate stream in sink j
AA	annualisation factor

6.3. Continuous Variables

$F^s_{i,j}$	allocated flowrate between sources i and sinks j
$F^d_{i,n}$	allocated flowrate between sources i and node n
$F^p_{q,j}$	flowrate of the permeate stream from regenerators q to sinks j
$F^r_{q,j}$	flowrate of the retentate stream from regenerators q to sinks j
F^a_n	flowrate of streams from re node n
P^a_n	pressure of streams leaving node n
P^i_n	pressure of an inlet stream to an energy recovery turbine from node n
P^o_n	pressure of an outlet stream from an energy recovery turbine from node n
P^r_q	pressure of a retentate stream from regenerator q
FW	freshwater flowrate
WW	wastewater flowrate

6.4. Binary Variables

$$y^{p}_{q,j} = \begin{cases} 1 \leftarrow \text{if piping exits between the permeate streams and sink } j \\ 0 \leftarrow \text{otherwise} \end{cases}$$

$$y^{r}_{q,j} = \begin{cases} 1 \leftarrow \text{if piping exits between the retentate streams and sink } j \\ 0 \leftarrow \text{otherwise} \end{cases}$$

$$y_{i,j} \begin{cases} 1 \leftarrow \text{if piping exits between source } i \text{ and sink } j \\ 0 \leftarrow \text{otherwise} \end{cases}$$

$$y^{d}_{i,n} \begin{cases} 1 \leftarrow \text{if piping exits between source } i \text{ and node } n \\ 0 \leftarrow \text{otherwise} \end{cases}$$

6.5. Integer Variables

N^{s}_{q} the number of hollow fiber modules of regenerator q

7. References

[1] S. Bandyopadhyay and C. Cormos, "Water management in process industries incorporating regeneration and recycle through a single treatment unit," *Industrial and Engineering Chemistry Research ,* vol. 47, pp. 1111-1119, 2008.

[2] C. Khor, D. Foo, M. El-Halwagi, R. Tan and N. Shah, "A superstructure optimization approach for membrane separation-based water regeneration networks synthesis with detailed nonlinear mechanistic reverse osmosis model," *Industrial and Engineering Chemistry Research ,* vol. 50, pp. 13444-13456, 2011.

[3] A. Alva-Argáez, A. Kokossis and R. Smith , "Wastewater minimisation of industrial systems using an integrated approach," *Computers and Chemical Engineering,* vol. 22, pp. 741-744, 1998.

[4] C. Khor, B. Chachuat and N. Shah, "A superstructure optimization approach for water network synthesis with membrane separation-based regenerators," *Computers and Chemical Engineering ,* vol. 42, pp. 48-63, 2012 .

[5] R. Tan, D. Ng, D. Foo and K. Aviso, "A superstructure model for the sybthesis of single-contaminant water network with partitioning regenerators," *Process Safety and Environmental Protection,* vol. 87, no. 3, pp. 197-205, 2009.

[6] C. Khor , B. Chachuat and N. Shah, "Optimization of water network synthesis for single-site and continous processes: milstones, challenges, and future directions," *Industrial & Engineering Chemistry Research ,* vol. 53, pp. 10257-10275, 2014.

[7] M. El-Halwagi, "Synthesis of reverse osmosis networks for waste reduction," *American Institute of Chemical Engineers,* pp. 1185-1198, 1992.

[8] F. Evangelista, "A short-cut method for the design of reverse-osmosis desalination plants," *Industrial & Engineering Chemistry Process Design and Development,* vol. 24, p. 211, 1985.

[9] Y. Wang and R. Smith, "Wastewater minimization," *Chemical Engineering Science ,* vol. 49, no. 7, pp. 981-1006, 1994.

[10] W. Kuo and R. Smith, "Effluent treatment system design," *Chemical Engineering Science ,* vol. 52, pp. 4273-4290, 1997.

[11] N. Takama, T. Kuriyama , K. Shiroko and T. Umeda , "Optimal water allocation in a petroleum refinery," *Computers and Chemical Engineering,* vol. 4, pp. 251-258, 1980.

Effect of Grid Generation in Numerical Simulation of Spray Granulation Tower

Liuhai Feng, Jiangyun Wang, Juan Wang, Fan Zhao, Yu Mao*
College of Chemical Engineering, China University of Petroleum, Beijing, P.R. China.

Abstract

The nozzle is the core equipment in the spray granulation tower. The separation efficiency of the Tower can be improved by adjusting the swirl intensity. However, the declining nozzle in the spray granulation tower increases the difficulty of geometric modeling. Researchers simplified the structure by establishing a vertical jet model with specific boundary conditions. To ensure the same flow rate with no-simplified nozzle, the researchers often increased the jet velocity or jet area, which often led to a deviation between the simulation result and actual result. To compare the different simulation result between the simplified nozzle and no-simplified nozzle, this paper researched the flow distribution in the spray granulation tower with arraying nozzles through the three-dimensioned numerical by numerical simulation and experiments respectively. The comparison of simulation and experimental result indicates that flow behavior and velocity distribution are difference between different grids. There is a longer central jet area and smaller atomization angle in no-simplified nozzle structure, which can weaken the numerical dissipation resulted from the inconsistency between flow direction and gridding direction near the nozzle entry. The simulation result of no-simplified spray granulation tower agrees well with experimental result. Therefore, it can obtain flow feature more accurately using no-simplified nozzle spray granulation tower.

Introduction

Due to the decreasing reserves and increasing weight of crude oil, the processing of heavy oil can effectively alleviate the serious energy crisis. Coupled with serious environmental problem, the process technology of heavy oil is obliged to be upgraded with higher efficiency. Heavy oil deep stage separation[1, 2] is a technique using supercritical solvent to extract asphalt which is hard to deal with in heavy oil. The processed oil is the important feedstock component of hydro-treating. The extracted materials are separated of solid solute from gas solvent in spray granulation tower. To improve the separation efficiency, it is necessary to study the flow behavior in the spray granulation tower. During the industrial oil production process, the fluid in the spray tower is the superficial fluid with drastic flash transition, so the flow process in the tower is extremely complicated. The experimental measurement is conducted under the atmospheric pressure and ambient temperature conditions, illustrated in Fig1.

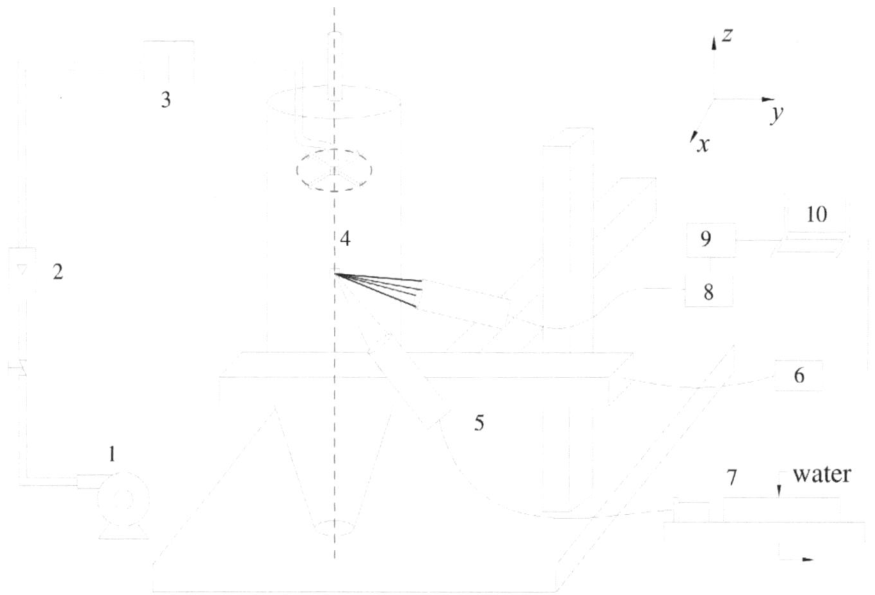

(1) compressor; (2) rotameter; (3) tracer particle generator; (4) spray granulation tower; (5) 3-D traverse system; (6) coordinate manipulator; (7) laser; (8) converter; (9) signal processor; (10) computer.

Fig.1 Schematic diagram of spray granulation tower

Simulation condition

The declining nozzle in the spray granulation tower increases the difficulty of geometric modeling. Langrish[3] and Liu[4] simplified the declining nozzle by establishing a vertical jet model with setting specific boundary conditions, which is used to ensure the consistence of the jet angle and the rate of flow between simulation and experiment.

According to the formula:

$$\begin{aligned} Q &= Av && \cdots No-simplified \\ &= \frac{A}{\cos\alpha} v\cos\alpha && \cdots Variation\ on\ orifice\ area \\ &= A\frac{v}{\cos\alpha}\cos\alpha && \cdots Variation\ on\ jet\ velocity \end{aligned} \quad (1)$$

so the jet area or velocity should be increased by $1/\cos\alpha$ times in simplified nozzle , respectively.

A structured hexahedron grid is used to establish the three different CFD models. The major difference of the three models is the spatial position of the nozzle. The flow in the tower is defined as turbulence according the Reynolds number. There is a rotational flow field in the tower, so the anisotropic RSM model[5, 6] is chosen as the calculating model. The nozzles are defined as Velocity Inlets, where the velocity magnitude and the flux are specified according to the experiment. The flow at the outlet of the tower is fully developed, so an ambient pressure is given at the outlet. The walls are set as the no-slip boundary conditions, and the standard wall function is applied to determine the flow near walls. The computation is carried out by the commercial computer software, Fluent. The governing equations is discretized by control volume integral method, and the SIMPLE algorithm is implemented to deal with the coupling between pressure and velocity, The solution is considered to be converged when the residual of the continuity is less than 10^{-4}.

Result and discussion

In consideration of the influence caused by grid generation, the spray granulation tower with single nozzle is simulated first in this study. The angle between the nozzle and the vertical plane is 45°, the diameter of the nozzle is 4 mm, and the jet velocity is 10 m/s. The cloud chart of velocity distributions is shown in Fig.2. Comparing the red elliptical benchmark, it finds that the atomizing angle of the no-simplified nozzle is obviously smaller than that of in simplified nozzle. So the increase of jet area or the jet velocity can enlarge the angle of atomization, and the increase of velocity has a more obvious influence on the change of angle. Comparing the black rectangular benchmark, it finds that the jet zone of no-simplified nozzle is longer than that of the simplified nozzles, and the increase of jet area has obvious influence on the attenuation of velocity. The central velocity along the jet direction is shown in Fig3. In the three models, the velocity has a same trend when S > 0.08 m. The air velocity of no-simplified nozzle maintains stability with some distance and decays subsequently. However, the air velocity of simplified nozzle decays quickly. In conclusion, the simplified structure has a great influence on the simulation result.

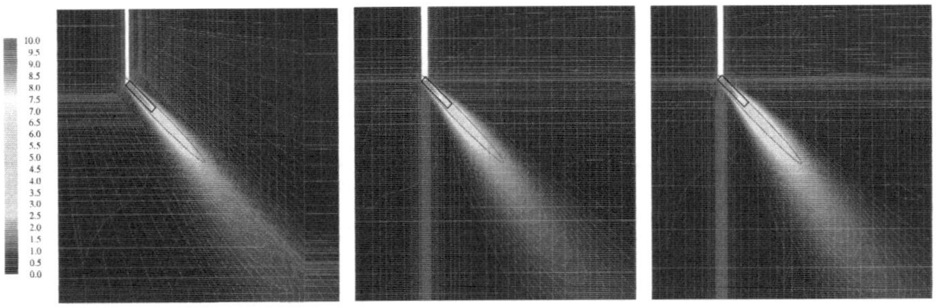

(a) No-simplified; (b) Variation on orifice area; (c) Variation on jet velocity

Fig.2 Comparison of different grid models

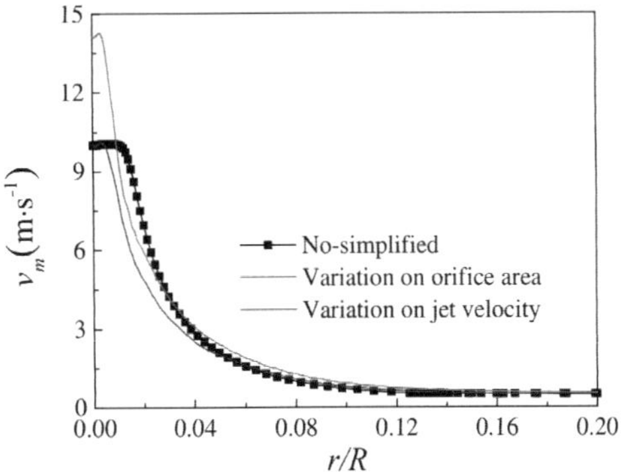

Fig.3 Variation of velocity along jet direction

Below the nozzle, the flow is influenced obviously by jet and the flow field is chaotic. The flow forms a stable rotational flow field as time goes on. So the flow field in the spray tower is divided into two parts: jet-dominated area and vortex stability region. The research in this article is mainly focus on the stability region, as shown in Fig.4. It can be seen that both the forced vortex in center and the free vortex outside can be acquired from simulation result in every structure. But the velocity distribution trend got from simulation result of simplified nozzle has some differences with the experimental result.

However, the simulation result of no-simplified nozzle coincides with experimental result better, including the peak value and trend of the tangential velocity. In conclusion, the no-simplified nozzle describes the flow process in the spray tower more accurately.

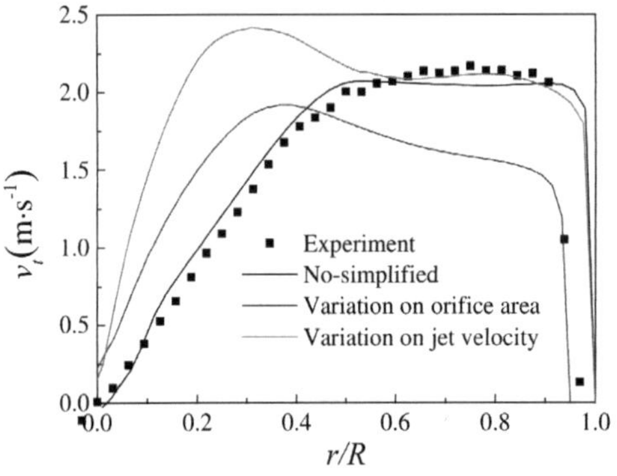

Fig.4 Comparison of velocity in simulation and experimental result

The simulation result is analyzed which is obtained from no-simplified nozzle spray granulation tower. Due to the interaction of jet and swirl flow, the flow field formed in SGT is complex, as shown in Fig.5. The flow in jet-dominated zone is greatly influenced by the jets. The high jet gas entrains the surrounding medium to the wall, and a spiral downward flow is formed as a result of the restriction of the wall. Besides, there are a lot of secondary vortices which the direction is oppose to the mainstream near the wall. The vertical eddy and secondary vortices are formed near the wall due to the combined effect of the jet and the entrainment. The flow pattern in the stable swirling zone is good. The effect of jet weakens gradually, resulting in the formation of the stable Rankine vortex swirl flow. The stable swirling zone is divided into two regions: an outer downward flow region and the center upward flow region. There exist a radial flow from outer to the center, causing turbulence in center region.

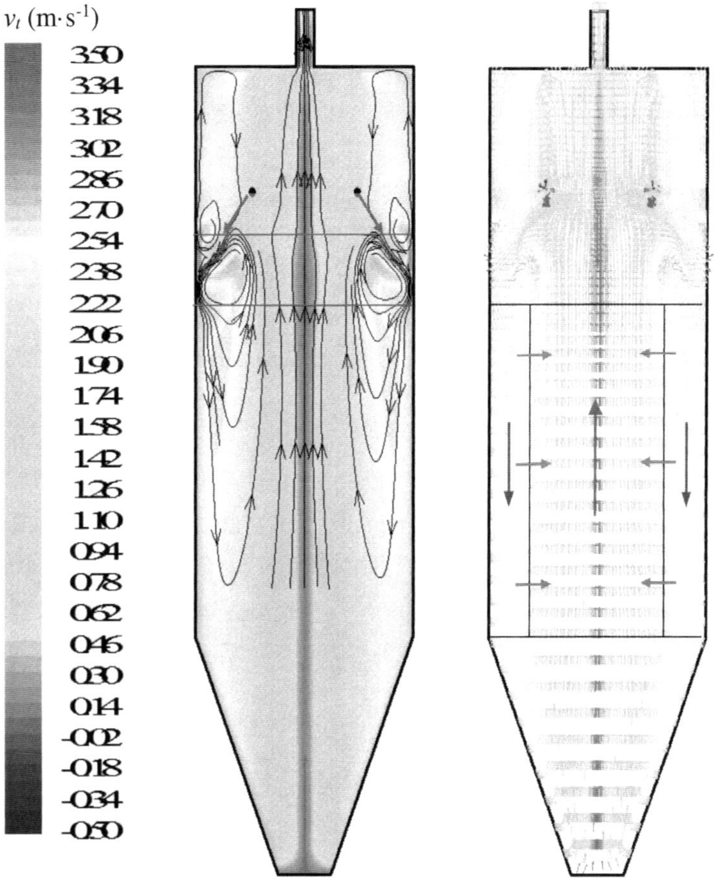

Fig.5 Distribution of tangential velocity and velocity vector in vertical section

Conclusions

The gas flow behavior in a jet-driven swirling flow is studied experimentally and numerically in this work. The flow dynamics, including flow development, velocity field and division characteristic, are revealed. The main conclusion can be summarized as follows:

(1) The no-simplified nozzle has a longer jet zone and smaller atomization angle, and the simulation results of the no-simplified nozzle has some difference with the experimental results. The simplified structure has some influence on the simulation results.
(2) Comparison of the simplified nozzle, the peak value and trend of the tangential velocity got from the no-simplified simulation coincide with the experimental result. The no-simplified structure of nozzle can weaken the numerical dissipation resulted from the inconsistency between flow direction and gridding direction near the outlet of the nozzle.
(3) There exists two regions in the spray tower: the jet-dominated area has a complex flow distribution, which has some vertical circulation and secondary vortex; the rotary area has a stable flow field, which has a flow distribution similar with the RanKine Vortex.

References

1. Shi T P, Hu Y X, Xu Z M, et al. Characterizing petroleum vacuum residue by supercritical fluid extraction and fractionation[J]. *Industrial & engineering chemistry research*, 1997, 36(9): 3988-3992.
2. Zhao S Q, Xu C M, Wang R A, et al. Deep separation method and processing system for the separation of heavy oil through granulation of coupled post-extraction asphalt residue [P]. US Patent, 7597794 B2, 2009.
3. Langrish T A G, Williams J, Fletcher D F. Simulation of the effects of inlet swirl on gas flow patterns in a pilot-scale spray dryer[J]. *Chemical Engineering Research and Design*, 2004, 82(7): 821-833.
4. Liu M, Mao Y, Wang J, et al. Effect of swirl on hydrodynamics and separation performance of a spray granulation tower with array nozzles [J]. *Powder Technology*, 2012, 227: 61-66.
5. Liu F, Chen J, Zhang A, et al. Performance and flow behavior of four identical parallel cyclones [J]. *Separation and Purification Technology*, 2014, 134: 147-157.
6. Zhao B, Su Y, Zhang J. Simulation of gas flow pattern and separation efficiency in cyclone with conventional single and spiral double inlet configuration [J]. *Chemical Engineering Research and Design*, 2006, 84(12): 1158-1165.

13 Ways Through a Firewall
What You Don't Know *Will* Hurt You

Compensating Measures

Attack Type	2FACT	ENC	RULES	HOST	NET	SUPD	UGW
1) Phishing / trojan / drive-by-download – victim pulls attack through firewall	0	0	2	1	1	1	2
2) Social engineering – steal a password / keystroke logger / shoulder surf	2	0	0	0	0	0	2
3) Compromise domain controller – create control system or firewall account	1	0	2	0	0	0	2
4) Attack exposed servers – SQL injection / DOS / buf-overflow / default pwds	0	1	1	1	1	1	2
5) Attack exposed clients – compromised web servers / file servers / data svrs	0	0	2	1	1	1	2
6) Session hijacking – MIM / steal HTTP cookies / command injection	0	2	1	0	1	0	2
7) Piggy-back on VPN – split tunneling / malware propagation	1	1	2	1	1	1	2
8) Firewall vulnerabilities – bugs / zero-days / default password / design vulns	0	0	0	0	1	1	2
9) Errors and omissions – bad firewall configurations / IT reaches through fwalls	1	1	1	1	1	1	2
10) Forge an IP address – firewall rules are IP-based	1	1	0	1	1	1	2
11) Bypass network perimeter – rogue cabling / wireless / cell phones / dial-up	1	1	0	1	1	1	0
12) Physical access to firewall – administrator ports / no password / modify hw	0	0	0	0	0	0	0
13) Sneakernet – removable media / untrusted laptops	0	0	0	1	0	1	0
Total Score:	7	7	11	8	9	9	20

Photo: Red Tiger Security

Abbr.	Compensating Measure
2FACT	2-factor authentication
ENC	Encryption, cryptographic authentication
RULES	Better firewall rules
Host	Host intrusion detection / prevention systems & SIEMs
Network	Network intrusion detection / prevention systems & SIEMs
SECUPD	Security updates / patch programs
UGW	Unidirectional security gateways

Grade	Description
2	Blocks essentially all attacks in this class
1	Blocks some attacks in this class
0	Not effective at blocking this class of attacks

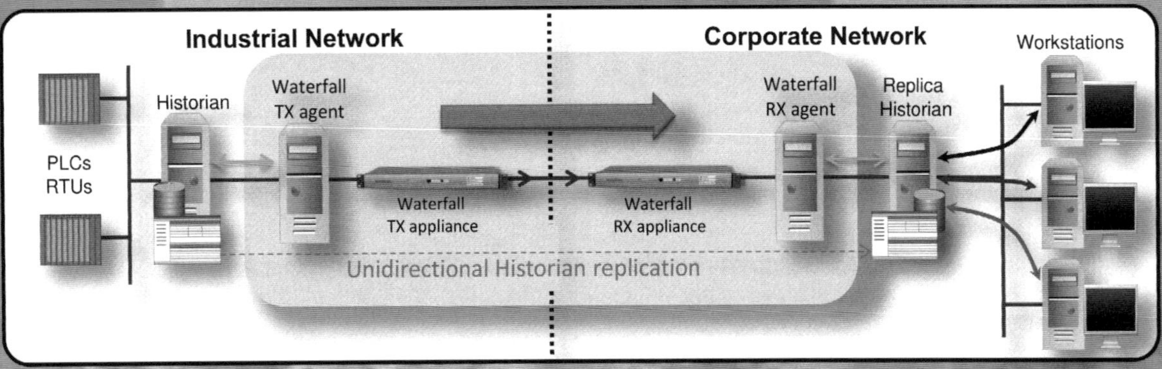

Unidirectional Security Gateways: Absolute Protection From External Network Attacks

WATERFALL®

Stronger Than Firewalls

400436 Optimisation of Refinery Diesel Blending

Abstract:

OPTIMISATION OF REFINERY DIESEL BLENDING

Shixun Jiang, Nan Zhang

Centre for process integration, University of Manchester

Oxford Road, Manchester M13 9PL, UK

In recent years, modern refineries have been confronted with higher prices of petroleum, more stringent environment standards and stricter requirements of quality specifications. To achieve the objective of maximizing the profit, it brings great challenge toward planning and scheduling for refinery operators.

Diesel, one of the main petroleum products, has a growing market proportion because of its higher energy utilization ratio and therefore lower carbon dioxide emissions compared with gasoline. Diesel is widely used in the industry and transportation. Only high quality diesel product can survive in the more and more competitive market. Refiners haven't stopped improving the methodology in diesel production and management.

To maximize the profit or minimize the cost is the main driving force in refinery diesel management. It is difficult to achieve the optimal recipe towards the site level as an optimal recipe for one single process may be suboptimal for other processes or even infeasible for the overall refinery. A multi-period planning model for refineries, which was based on the decomposition method, was proposed by Dave in 2004, to optimise decisions at different time intervals. Furthermore to optimise the recipe, refinery operators have to deal with volatile prices of crude oil and products, fluctuant market demand, uncertainty of feedstock arrivals and other abrupt changes. Sourabh (2008) demonstrated an "Operating Window" algorithm, which decomposed an MINLP problem into an NLP model and an MILP model.

In this work, a model for management of refinery diesel streams has been developed to optimize the diesel production of a refinery. A modified property estimation model for diesel, which is more accurate than traditional linear models, is employed for off-line diesel blending processes. Intricate environment regulations and product specifications, such as cetane number, sulphur content and pour point, etc. are taken into account. A novel scheduling method, which involves an MINLP model, is optimized by a robust solving algorithm. By adopting the method, the optimal solution of schedule a refinery diesel streams can be reached. Case studies have been carried out to demonstrate the effectiveness of the developed method.

References

Dhaval Dave, 2004. Planning and scheduling of refinery operations. Ph.D Thesis, University of Manchester, Faculty of Engineering and Physical Sciences.

Sourabh Gupta., 2008. Planning and scheduling of refinery operations. Ph.D. Thesis, University of Manchester, Faculty of Engineering and Physical Sciences.

400598 Electrochemical Reactor Modeling and the Determination of Minimum Operating Parameters for Conductive Paint Antifouling Systems

Abstract:

Protection from fouling organisms (biofouling) is a key function of coatings on marine hydrokinetic (MHK) devices. Accumulation of biomass on devices risks performance degradation and possible failure. Additionally, artificial structures in the marine environment have been shown to encourage settlement of invasive species, necessitating application of mitigation strategies.[1] At the present there are no proven reliable options for prevention of fouling in the marine environment with applications to match device lifetimes expected to be on the order of 10-20 years. Removal, cleaning, and re-application of short-lived antifouling paints is an expensive prospect. Electrified conductive paints held at sufficient electric potential have been shown to be effective in the prevention of marine biofouling, and may be a viable solution for long term antifouling performance[2] . The present work seeks to extend the understanding of performance variables for such coatings. Modeling is used to predict chemical species concentrations in fluid flow over a charged flat plate. Minimum operating conditions required for inhibition of algal growth in filtered seawater are then determined experimentally in a flow cell with applicable geometry to elucidate the mechanism of action.

The mechanism of action has been proposed variously to be direct electron transport[2], electromagnetic field repulsion[3], and electrochemical reaction[4, 5]. Evidence suggests that electrochemical oxidation of chloride and bromide present in seawater produces strong oxidizing agents which sanitize the positively charged anodic surface. The negatively charged cathodic surface is likely driving the hydrogen evolution reaction, which increases the local pH to intolerable levels for marine fouling organisms. Comparison of the experimentally determined minimum potential needed for biofouling protection with predicted equilibrium potentials for each reaction and Tafel plots in filtered seawater support this hypothesis.

[1] L. H. Hedge and E. L. Johnston, "Propagule pressure determines recruitment from a commercial shipping pier," *Biofouling,* vol. 28, pp. 73-85, 2012/01/01 2012.

[2] T. Matsunaga, T. Nakayama, H. Wake, M. Takahashi, M. Okochi, and N. Nakamura, "Prevention of marine biofouling using a conductive paint electrode,"*Biotechnol. Bioeng.,* vol. 59, pp. 374-378, // 1998.

[3] T. Nakayama, H. Wake, K. Ozawa, N. Nakamura, and T. Matsunaga, "Electrochemical prevention of marine biofouling on a novel titanium-nitride-coated plate formed by radio-frequency arc spraying,"*Appl. Microbiol. Biotechnol.,* vol. 50, pp. 502-508, // 1998.

[4] R. E. Perez-Roa, M. A. Anderson, D. Rittschof, C. G. Hunt, and D. R. Noguera, "Involvement of reactive oxygen species in the electrochemical inhibition of barnacle (Amphibalanus amphitrite) settlement," *Biofouling,* vol. 25, pp. 563-71, // 2009.

[5] J.-R. Huang, W.-T. Lin, R. Huang, C.-Y. Lin, and J.-K. Wu, "Marine biofouling inhibition by polyurethane conductive coatings used for fishing net," *J. Coat. Technol. Res.,* vol. 7, pp. 111-117, // 2010.

[6] *Paint and Coating Testing Manual : 15th edition of the Gardner-Sward handbook*: ASTM International, 2012.

400962 Advanced 3-Phase Flash Computations in the API Tech Data Book 9 Software

Abstract:

Flash computations and the physical properties of each phase are vitally important for accurately sizing process equipment. The 3-phase flash/mixture calculations in the API Tech Data Book 9th Edition Software provide highly accurate flash results for hydrocarbon mixtures. This is due to the incorporation of accurately tuned binary interaction parameters and the use of industry proven equation of state methods for a very wide range of components and operating conditions. This allows for accurate flash calculations to be done by any chemical engineer working in the oil and gas industry for most systems of interest without any indepth knowledge of thermodynamics and thermophysical properties. Comparison of results by the API Tech Data Committee against 3 major simulaiton software products showed superior results in 5 areas important to accurately sizing process equipment. The software interface uses a modern XAML design that does not require a flowsheet to be created so that chemical engineers not experienced with process simulation software can quickly become proficeint in its use.

CD-adapco

2015 AIChE Spring Meeting &
11th Global Congress on Process Safety
April 26 - 30, 2015 • Hilton Austin • Austin, TX
www.aiche.org/spring

A CFD Model-Based Optimization of a Process Burner Geometry

Krishnamoorthy, N., Halliday, N., Dewan, Y., and Aglave, R.,

Problem Description

CD-adapco

Design optimization of a single flat flame burner using STAR-CCM+ and Optimate+

Optimate+ is an easy to use plugin that utilizes SHERPA algorithm for design space exploration. It is:

- Hybrid – blend of search strategies, global and local search
- Adaptive – adapts to design space, searches simple and complex spaces

Objectives:

- Minimize: CO and NOx relative to the baseline design
- Explore: The dependence of Flame Height and Emissions on burner design parameters
  - Influence of tile tilt angle
  - Influence of fuel port spacing

Geometry and Operating Conditions

- **4 Fuel ports**
- **Fuel and oxidizer at STP**
- **Burner Heat Release = 0.88 MW**
- **Excess Air Supplied = 10%**

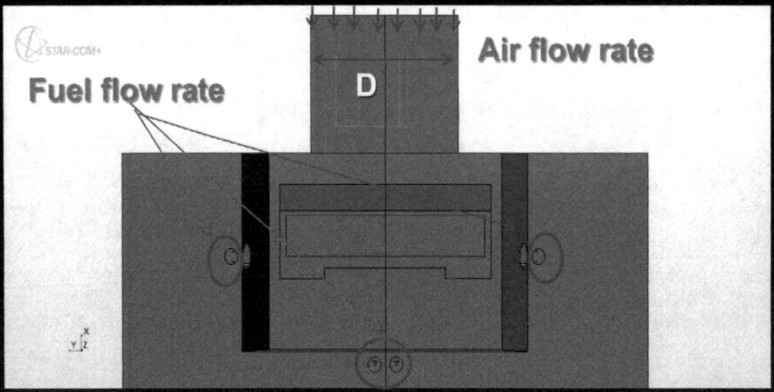

Numerical Model and Optimization Methodology

Numerical Model Selection in STAR-CCM+

- Three Dimensional
- Half Burner : Symmetry Condition
- SST K-Omega Turbulence Model
- PPDF Flamelet Combustion Model
- Thermal NOx: NOx PPDF Flamelet
- Discrete Ordinates Radiation Model
  - Weighted Sum of Gray Gases

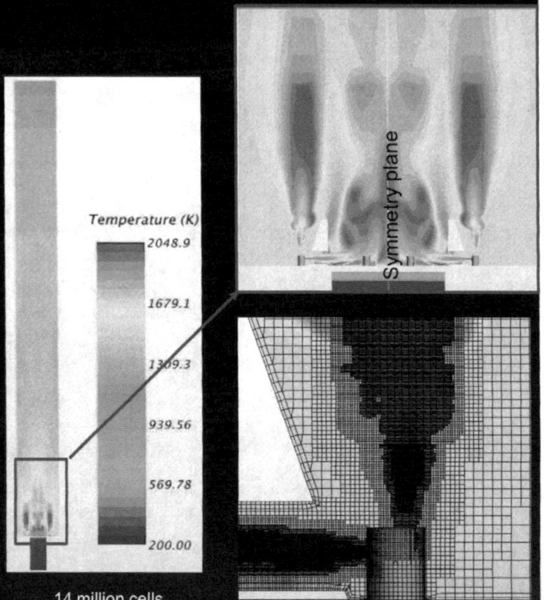

14 million cells

Fuel jet resolution using table based refinement

Optimization Variables

CD-adapco

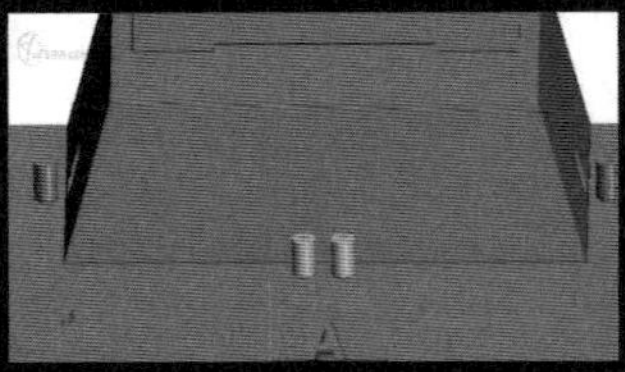

A: Primary Injector Spacing
Minimum: 25 mm
Maximum: 250 mm
Intervals: 50

B: Burner Tile Angle
Minimum: 1
Maximum: 35
Intervals : 10

C: Primary Inj. Nozzle Angle
Minimum: 1
Maximum: 20
Intervals : 10

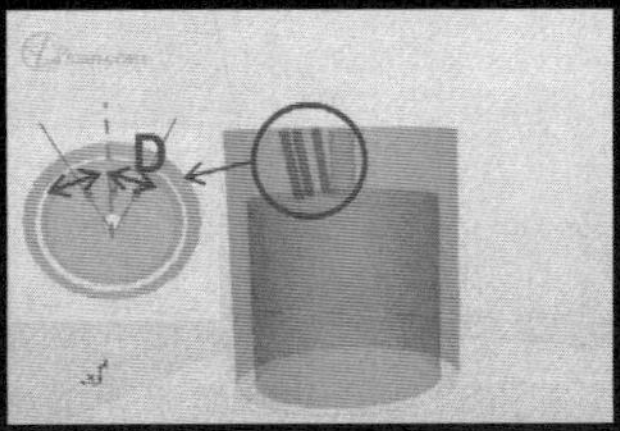

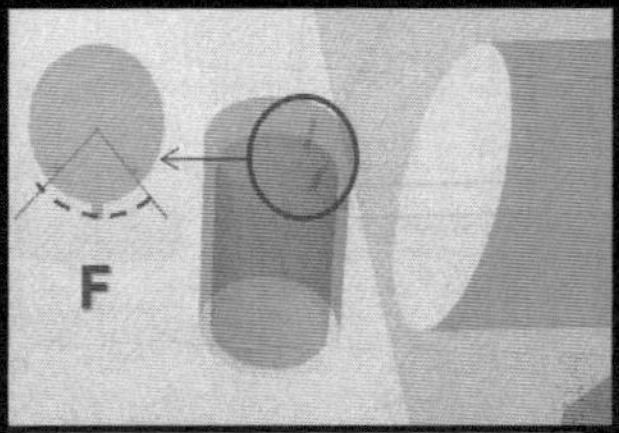

D: Primary ports angle spacing in circular pattern
Minimum: 5
Maximum: 45
Intervals : 11

E: Side Inj. nozzle angle
Minimum: 1
Maximum: 30
Intervals : 10

F: Side nozzle angle spacing in circular pattern
Minimum: -120
Maximum: 60
Intervals : 50

Burner Optimization: Responses and Constraints

CD-adapco

Responses

- **Flame Height**
  - Defined by iso-contour of CO = 1200 ppm
- **Flame Volume**
  - Volume occupied by iso-contour of CO = 1200 ppm
- **CO concentration at Exit**
  - Mass flow averaged CO mass fraction at outlet
- **NOx concentration at Exit**
  - Mass flow averaged NOx mass fraction at outlet
- **Temperature at Exit**
  - Mass flow averaged Temperature at outlet

Constraints

- **Volume Average Temperature**
  - In the firebox > 1350 K
- **Flame height**
  - > 1.695 m (30% of baseline)

Pareto Optimization Results

- **Minimize CO and NOx emissions at exit**
  - CO and NOx are competing objectives

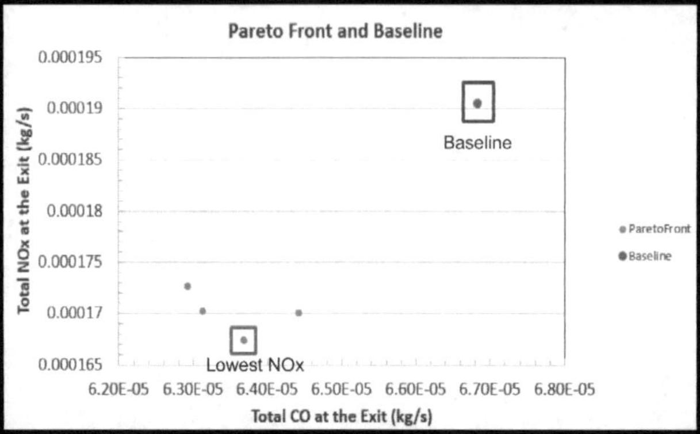

- **Total of 40 designs were run**
- **15 designs did not meet the constraint. These were rejected**

Best design minimizes CO by 4.7% and NOx by 12% from the baseline

Influence of Tile Tilt Angle

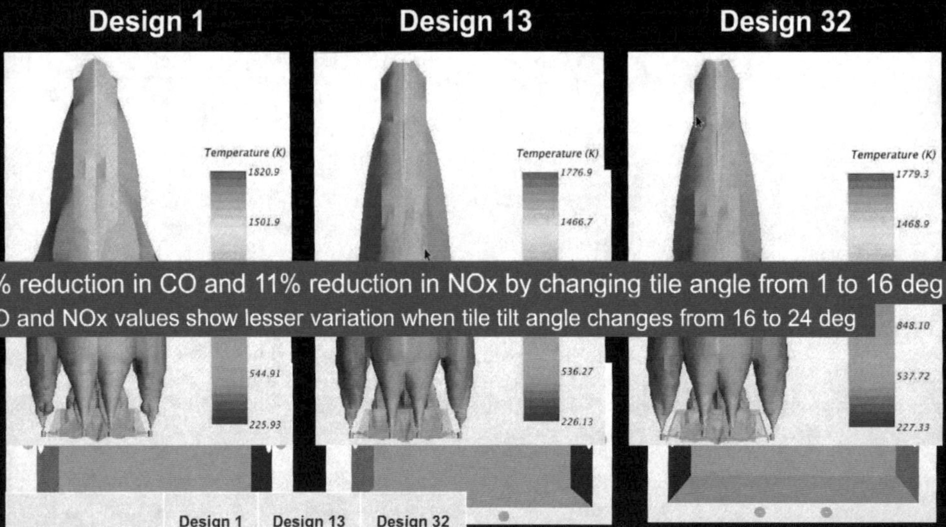

4% reduction in CO and 11% reduction in NOx by changing tile angle from 1 to 16 deg

CO and NOx values show lesser variation when tile tilt angle changes from 16 to 24 deg

	Design 1	Design 13	Design 32
Flame Height (m)	2.422207	2.342207	2.42221E+00
Flame Volume (m^3)	2.15E-01	1.86E-01	1.84E-01

3% reduction in flame height for Design 13
Similar flame height to baseline for Design 32
~14% reduction in flame volume for Designs 13 and 32

Influence of Fuel Injector Spacing

CD-adapco

Design 17
Design 17
Injector Spacing 25 mm

Design 1
Design 1
Injector Spacing 88 mm

Design 28
Design 28
Injector Spacing 250 mm

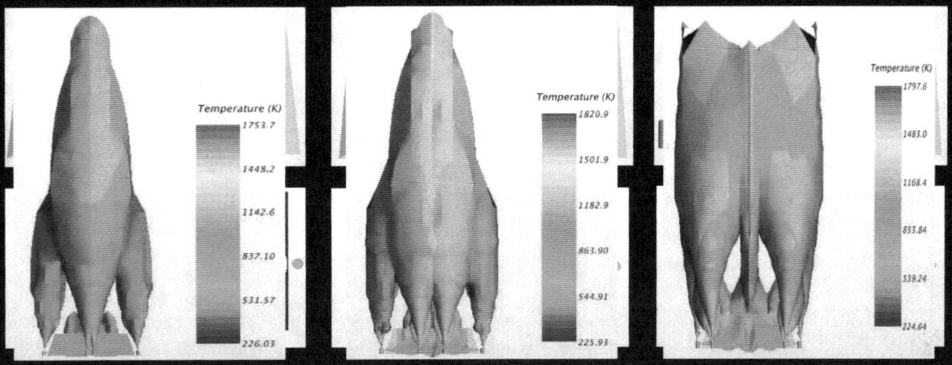

Flame height and Flame volume for Design17 higher than Design 1. Design 17 has 8% higher NOx
5% reduction in CO and 3.4% reduction in NOx for Design 28. But Flame height reduced by 28%

Summary of Design Optimization

CD-adapco

- **Exploration:**
  - Tile tilt angle change from 1 deg to 16 deg shows reduction in both CO and NOx .
  - Reduction in overall flame height and volume seen with increasing tile angle
  - Fuel injector spacing affects flame height and flame volume. This in turn affects the emission values
- **Optimization**
  - Best design that results is minimum CO and NOx values is Design 32.
  - CO reduced by 4% from baseline
  - Simultaneous NO reduction by 11% from baseline

401341 Modeling, Optimisation and Control of Solar Absorption Air-Conditioning System

Abstract:

Solar thermal energy can provide a benign and environmentally friendly heat source for driving a vapour absorption refrigeration system of unit capacity using R 717 (NH3) and water as the working fluids. The last few decades witness an upsurge in energy demand, both at the domestic and industrial cooling and heating demand. Traditionally, heating, ventilation and air-conditioning (HVAC) performance and design are based on construction of prototypes in a wind tunnel. This approach is not only time consuming but expensive. In this study, a model-based approach for the modelling, optimisation and control of a solar vapour absorption air-conditioning system is presented. A model-in-loop and hardware-in-loop model was also develop to validate the control algorithm.

401412 New Development on Hoist Scheduling for Multi-Stage Material Handling

Abstract:

New Development on Hoist Scheduling for Multi-stage Material Handling

HonglinQu, Chuanyu Zhao and QiangXu*

Dan F. Smith Department of Chemical Engineering

LamarUniversity, Beaumont, TX77710, USA

Abstract

Multi-stage material handling (MSMH) processes are widely used in industries for the manufacture of various workpieces and products (jobs) in large scale. Generally, a hoist, as control device, will handle all jobs by following a preset movement schedule, including joblifting, releasing and moving along a production line. Lots of engineers have studied the feasibility and optimization of hoist movement schedules. However, the productivity of an MSMH process not only depends on the hoist schedule, but also relies on the production line arrangement, such as the spatial allocation of processing units. During production, hoist will travel forth and back to pick up job from its current unit and release it into the next one, following its processing requirement. In this way, a proper arrangement of unit position would dramatically shorten the total moving time and improve the processing capability. As far as we know, no work has been done in this field. Therefore, we try to arrange the unit allocation in a more compact way to optimize the hoist schedule.

In this paper, a two-dimensional (2D) matrix arrangement of processing unit has been proposed to replace original one-dimensional (1D) line arrangement for the optimization of productivity, and a mixed integer linear programming (MILP) model has been developed to solve this problem. The major advantage of this study is that the compacted allocation provides shorter paths for hoist to move, so the productivity could be improved significantly. The efficacy of the proposed methodology is demonstrated by a case study. The comparisons between 1D and 2D production line arrangements have also been conducted.

401413 Simultaneous Scheduling of Front End Crude Transfer and Refinery Processing (Electronic Poster Presentation)

Abstract:

Simultaneous Scheduling of Front End Crude Transfer and Refinery Processing

Jialin Xu, Shujing Zhang, Jian Zhang, and Qiang Xu

Dan F. Smith Department of Chemical Engineering

Lamar University, Beaumont, TX 77710

Abstract

Scheduling and planning of crude oil is very important to a petroleum refinery due to the potential realization of large cost savings. It is also the one of most challenging problems for refinery operations. In this paper, a new continuous crude scheduling model has been proposed to simulate and optimize the front-end and refinery crude-oil operations. This scheduling problem involves crude oil unloading from vessels to the crude storage tanks at onshore berths, transfer of crude from these tanks to the charging tanks, charging crude distillation units with crude mixes from the charging tanks and further processing of crude inside the refinery, which includes crude distillation, cracking, coking, reforming, hydrotreating, product blending and component recovery.

The objective of the proposed crude scheduling problem is to maximize the total operational profit; meanwhile, it should meet inventory and production demands, and try to maintain refinery plant's normal operations under uncertainties. The scheduling model is a large-scale mixed integer nonlinear programming problem(MINLP). An outer-approximation based decomposition method is implemented to obtain the optimum solution of the developed MINLP scheduling model.

401527 An Optimization Framework for the Day-Ahead Scheduling of District Cooling Systems

Abstract:

The main goal of most district cooling systems is to centrally produce and distribute chilled water at a minimum cost while satisfying demand. Chilled water is produced by production plants and distributed to buildings equipped with energy transfer stations (sub-stations). After the heat transfer takes place, water is returned to the production plants where the chilled water is regenerated. District cooling is essential for sustainable economic and urban development. The reason for this is that by centralizing the chilled water generation, district cooling systems can increase system efficiency, reduce air pollution, decrease emissions of refrigerants, and manage electricity demand on-site and to the electric grid as a whole.

Chillers are the main components of the chilled water production plants. The production cost of chilled water plants that use electrically powered chillers is deeply sensitive to electricity price variability [4]. To minimize the production cost, thermal energy storages are often used [1,3]. The main idea is to store chilled water in the thermal energy storage when the electricity price is low and discharge it when the electricity price is high. Although this operating rule may seem trivial to implement, multi-modal electricity price curves, thermal energy storages with limited capacity as well as multi-period constraints on chillers present a challenge to the operators who aim at developing the day-ahead operating schedule of the cooling system at a minimum cost.

In this work we propose an optimization framework that deals with this situation. We show that by combining the latest results on piece-wise linearization techniques [2] with disjunctive programming strategies [5] we are able to generate an optimization model which can be solved fast and efficiently as a mixed-integer problem. We describe how this framework is implemented and successfully used to calculate the day-ahead optimal schedule of the thermal energy storage and chillers at the Thermal Energy Corporation, a district energy company located in Houston, TX.

[1] W. Bahnfleth and W. Joyce. 1994. Energy use in a district cooling system with stratified chilled water storage. ASHRAE Transactions. Volume 100, pp 1767-1778

[2] B. Geißler, A. Martin, A. Morsi, L. Schewe. 2012. Using Piecewise Linear Functions for Solving MINLPs. The IMA Volumes in Mathematics and its Applications. Volume 154, pp 287-314

[3] W. J. Cole, K. M. Powell and T. F. Edgar. 2012. Optimization and advanced control of thermal energy storage systems. Chemical Engineering Reviews. Volume 28, pp 81–99

[4] K. M. Powell, W. J. Cole, U. F. Ekarika, T. F. Edgar. 2013. Optimal chiller loading in a district cooling system with thermal energy storage. Energy. Volume 50, pp 445–453

[5] J. P. Ruiz, J. Wang and C. Liu. 2013. Logic-based outer-approximation method for the security constrained unit commitment problem. IET Generation, Transmission & Distribution. Volume 7, pp 1210-1218

401529 Optimized Control Strategy of Distillation Column Under Startup and Shutdown

Abstract:

The startup and shutdown operations of a distillation column involve complex heat and mass transfer phenomena in different transient time periods. Dynamic simulation can be used to study and analyze these dynamic behaviors. Based on dynamic simulations, column operating procedures and control strategies can be improved to guarantee the operations flexible and reliable. And operation time and energy savings have an appreciable economic significance.

In this study, rigorous dynamic simulations are employed to investigate the startup and shutdown of C2 splitter in an ethylene plant. First, in the base case, the dynamic model for the column is built with normal control strategy. All the operating procedures are programmed and automated in one rigorous simulation model. Based on the results in the base case, the control strategy is improved in the optimized case. Second, dynamic model for C2 splitter is built with optimized control strategy. Split range control and low select control are applied in this optimized control strategy. The automation for start-up and shutdown operation is enhanced. In summary, dynamic simulations performed in this study will help identify operating procedures and control strategies to start up or shut down a distillation column quickly and safely, so as to save plant raw material and energy, reduce operational risks, and thus enhance the industrial sustainability.

401556 On-Line Optimal Control of Molecular Weight Distribution in Batch and Semi-Batch Free Radical Polymerization Processes

Abstract:

Polymer molar mass distribution is one of the most important polymerization control variables that affect many polymers' characteristics such as strength and thermal stability. Although there have been full feedback control of reactors on small molecules, polymerization reaction control is less advanced due to lack of understanding of the dynamics of the process, inadequate in situ sensors and nonlinearity in the behavior of the reaction for instance gel affect that influences kinetic parameters during the reactions.

This contribution presents a complete real time feedback framework to control the conversion kinetics and polymer chain length distribution during the free radical batch polymerization operation. In the first step the process is conducted in batch mode where a detailed polymerization model is applied along with the method of finite molecular weight moments to determine a sequence of reactor temperature set points which lead to the desired molecular weight distribution. The optimization of the model is performed using the gOPT function in gPROMS. In order to have a better controllability on the process the system is also simulated in semi-batch mode where reagents are fed into the reactor during the reaction. The formalism in this case combines kinetics with time dependent processes related to flows of material into and out of the reactor. The objective is to find the monomer (initiator) flow rate to control the average molecular weight and the temperature is adjusted to manipulate the polydispersity. To validate the model and assess the merits of the proposed strategy, free radical polymerization of methyl methacrylate is carried out using 2,2'-azobis(2-methylbutanenitrile) as initiator. Automatic Continuous Online Monitoring of Polymerization reactions (ACOMP) is used to follow the conversion and evolution of the average mass distribution, and multi-detector SEC is applied to cross-check results and measure full distributions of end products.

401655 DNA Hybridization on Silica Surface for Gene Chip Design Studied with Molecular Dynamics Simulations

Abstract:

Nucleic acid hybridization behavior has attracted research attention due to their unique pairing interactions. The short nucleic acid probes such as miRNA or 16s regions of microbial genomes are being applied as biomarker for the detection of various diseases, such as cardiovascular or infectious diseases. To facilitate our experimental design of gene chip for early diagnostic purpose, we performed a fully atomistic Molecular Dynamics simulation to study DNA hybridization behavior on α-cristobalite (101) surfaces, which is partially hydroxylated and also chemically modified with aldehyde chains. Drawn on the atomistic simulations, we investigated the morphology of the grafted DNA strands on the surface and the interactions between DNA complementary strands, as a function of the surface properties and solution environment (such as ion strength) for the purpose of enhancing hybridization efficiency. The current research will also lay the foundation for future research with the coarse-grained modeling.

401676 Structures, Dynamics, and Water Permeation Free Energy Across Bilayers of Lipid a and Its Analog Studied with Molecular Dynamics Simulation

Abstract:

Fundamental studies of the supramolecular layer structures, dynamics and water permeation free energy of hexa-acyl-chain Lipid A and its analogue of tetra-acyl chains would be useful for polymer membranes design for endotoxin removal in water treatment, drug delivery and other biotechnologies. In this work, we studied their supramolecular bilayer by using molecular dynamics simulations and efficient free energy computations. Our simulation accuracy was verified by the agreement between the bilayer structural properties (structure factor, bilayer thickness, and the area per lipid) and lateral diffusion coefficient in our simulation and experimental measurements. More importantly, our simulation for the first time illustrated hexagonal compact packing of the hydrocarbon acyl chains within a leaflet of Lipid A membrane (at 298 K and water content of 40 wt %), which is consistent with experiments. In contrast, Lipid A analogue is found with less ordered ripple structures at the same condition. Our study also demonstrated slower dynamics and larger and broader free energy barrier (23 kJ/mol) for water permeation for Lipid A, compared with that of Lipid A analogue. Moreover, the analysis of dynamics showed that highly hydrated hydrophilic diglucosamine backbone is structurally stable, whereas the interdigitated hydrophobic acyl chain tails inside the membrane with faster dynamics screen the aqueous environment from the lipid interior and also reinforce the membrane's structural stability.

401681 Lysozyme Adsorption and Desorption on the Azobenzene Self-Assembling Surfaces Studied with Molecular Dynamics Simulations

Abstract:

The reversible photosponsive isomerization of the azobenzene and its derivatives can control protein structure in an aqueous environment with the alternation of visible and UV lights for very promising applications in drug delivery and biomaterials. However, protein adsorption/desorption behavior on the azobenzene self-assembled monolayer surface (Azo-SAMs) is still unknown. In this study, atomistic molecular dynamics simulations were performed to study lysozyme adsorption and desorption on a silica surface grafted with Azo-SAMs. Protein structural arrangements (i.e., the shape and secondary structures), its mobility and the interactions with the surfaces, as a function of the azobenzene SAMs density and morphology, are systematically investigated.

401708 Improving Process Sustainability and Profitability for a Large U.S. Gray Iron Foundry (Electronic Poster Presentation)

Abstract:

Improving Process Sustainability and Profitability for a Large U.S. Gray Iron Foundry

Prashant Nagapurkar, Shyam Paudel and Joseph D. Smith, PhD.

Missouri University of Science and Technology, Rolla, Missouri, USA

Abstract

Energy savings and sustainability has been an important topic in many industrial processes. Limited energy resources and increasing electricity prices make foundry processes less competitive in today's world economy. For foundry processes, the energy costs generally represent 5-7% of the total sales. Most of the energy used in an foundry is directly related to melting and heat treatment operations. Integrating heat and power systems in a foundry through waste heat recovery techniques can lead to energy reutilization that improves process sustainability and profitability. The work presented here is an in-depth energy management analysis using Powersim^{TM} to identify potential energy savings through process optimization and advanced control strategies. A system dynamics model based on the interactions among the major energy intensive variables has been developed using Powersim^{TM} to estimate energy consumption. Our dynamic model estimates the monthly energy usage and its costs for a full calender year. The estimated energy consumption and its costs have been validated by comparison to the actual plant data obtained from the Mapleton foundry.

The average metal caster has a 2.4% pre-tax operating profit on sales. This profit is susceptible to fluctuating utility costs and process disturbances. Since foundry operates with very thin profit margins it is imperative to ensure that the foundry is running with efficient process control. Hence, advanced control strategies for the melters, dust collectors, fans and burners have been proposed and implemented in the Powersim^{TM} model for optimum energy usage. Our results show that these control strategies along with waste heat recovery techniques could save nearly 18% in energy savings and $1.8 million in costs annually for the Mapleton foundry.

400426 Synthesis of Efficient Multicomponent Separation Processes

Abstract:

Separation processes find application in a variety of industries like the chemical, petrochemical, pharmaceutical, gas separation and semiconductor industries. It is estimated that separation processes account for nearly 40-70% of chemical plant costs. One reason for the associated high costs is that the industry relies on the experience and intuition of a few expert engineers to build separation sequences for any given application. In this work, we present a systematic method to design a separation system that is up to 50% cost-effective and energy-efficient over existing systems. Our focus is on distillation configuration design. First, we enumerate all useful distillation sequences that are possible for separating a given mixture, and then incorporate them into an optimization framework which produces a rank-list of distillation sequences based on a desired criterion (energy consumption/overall cost/thermodynamic efficiency). Such a tool in the hands of an industrial practitioner will allow the person to focus on a certified small subset of useful configurations. Furthermore, we present some exciting recent progress made in process intensification for distillation, in areas of dividing wall columns and simultaneous heat and mass integration.

399546 An Extensive Comparison Study of Batch Process Monitoring Approaches

Abstract:

Batch processes play a central role in modern industrial plants, namely in the production of high valued products and fine chemicals, due to their added flexibility and scalability. The operation of batch processes is also intrinsically dynamic and non-stationary, bringing more degrees of freedom to their operation but increasing the complexity of the optimization and monitoring tasks. Regarding process monitoring, batch processes generate data with a 3-way structure, $\boldsymbol{X}$ (I x J x K), where I is the number of batches, J is the number of variables and K is the number of measurements made during a batch. One of the first methods proposed for addressing this problem is Multiway Principal Component Analysis (MPCA), which is based on the batch-wise unfolding of the 3-way data matrix followed by the application of a 2-way latent variable methodology, such as PCA or PLS. Further developments led to the variable-wise unfolding and 3-way approaches. More recently, dynamic methods were also introduced with the intent to explicitly model the process dynamics.

MPCA for batch process monitoring was first proposed by Nomikos and MacGregor [1]. The methodology starts by unfolding the 3-way data matrix into a 2-way matrix $\boldsymbol{X}$ (I x JK): vertical slices (I x J) are placed side by side such that the first slice corresponds to all variables at the first time point and subsequent slices, regarding subsequent times, are arranged to the right of the existing ones. The variables of the extended $\boldsymbol{X}$ matrix are then mean centered and their variance scaled to unity. This corresponds to a batch-wise normalization, which essentially removes the batch mean non-stationary trend. The remaining variation is then modeled by PCA. As a result of the unfolding procedure, during online monitoring it is necessary to predict future observations in order to compute the scores. To do so, several in-filling approaches are available such as: zero deviation, current deviation and projection data.

The class of 3-way approaches is based on the use of different trilinear decompositions of X (I x J x K) performed by several chemometric methods, such as PARAFAC and Tucker3. Tucker3 provides a decomposition of the 3-way array into three orthogonal loadings (A, B and C) and a core matrix (G) plus a residual matrix (E). The PARAFAC model is written in terms of the product of the three loading matrices (A, B and C) plus a matrix of residuals (E). PARAFAC differs from Tucker3, mainly because each of its three dimensions cannot be flexibly combined as occurs in the Tucker3 model, due to the nature of its core matrix. Furthermore, PARAFAC also requires that all loading matrices share the same number of components.

Regarding the class of dynamic methodologies, Chen and Liu [2] proposed a Batch Dynamic PCA (BDPCA) approach that extends the Dynamic PCA method of Ku, Storer [3] to batch processes. In this method an extended data matrix of time-shifted variables is constructed for each batch and used to compute an extended sample covariance matrix. A standard PCA model is then build based on the average sample covariance matrix.

Alternatively, instead of monitoring the measured observations, Choi, Morris [4] proposed the monitoring of their residuals with resort to an autoregressive PCA (ARPCA) method. Basically, this method fits a multivariate autoregressive model to the data and then uses it to compute one-step-ahead predictions of the current measurements. Afterwards, the residuals are determined and organized variable-wise prior PCA analysis.

Based on the BDPCA scheme, the application scope of Dynamic PCA with Decorrelated Residuals (DPCA-DR) proposed by Rato and Reis [5] can be easily adapted to the case of batch monitoring. To apply this approach a DPCA model is constructed as in the BDPCA methodology. However, in DPCA-DR the monitored quantities are the one-step-ahead predictions of the scores and observations, which are obtained through the application of missing data imputation techniques [6], assuming that the current observations are missing. As a result, the monitoring statistics present low autocorrelation levels and have a similar behavior to those of ARPCA but without the need of fitting a multivariate time series model.

As can be observed from the discussion above, a wide range of methods for online monitoring are currently available. However, only very few studies have been conducted to compare particular aspects of their relative performance. It is therefore the objective of this work to conduct an extensive study that contemplates the leading strategies for online batch process monitoring, in order to provide practitioners with a clear and updated vision of the advantages and disadvantages of each method.

In order to conduct the proposed task, normal and faulty batches generated from simulated batch processes were monitored by the above methods. The respective control limits were computed using current and leave-one-out approaches. For the residual statistics, different window sizes using k observations before and after the current residuals were also considered. The assessment of the studied monitoring methods, as well as the effects of the different in-fillings and control limits, was made taking into account two distinct dimensions of performance: detection strength and detection speed. Detection strength was evaluated through a performance index based on the receiver operation curve (ROC) of the methods in each trial, which provides a robust measure of their ability to discriminate among normal and faulty batches. Detection speed was assessed through the analysis of the Conditional Expected Delay versus Probability of False Alarms curve.

References

1. Nomikos, P. and J.F. MacGregor, *Monitoring batch processes using multiway principal component analysis.* AIChE Journal, 1994. **40**(8): p. 1361-1375.

2. Chen, J. and K.-C. Liu, *On-line batch process monitoring using dynamic PCA and dynamic PLS models.* Chemical Engineering Science, 2002. **57**(1): p. 63-75.

3. Ku, W., R.H. Storer, and C. Georgakis, *Disturbance detection and isolation by dynamic principal component analysis.* Chemometrics and Intelligent Laboratory Systems, 1995. **30**(1): p. 179-196.

4. Choi, S.W., J. Morris, and I.-B. Lee, *Dynamic model-based batch process monitoring.* Chemical Engineering

Science, 2008. **63**(3): p. 622-636.

5. Rato, T.J. and M.S. Reis, *Fault detection in the Tennessee Eastman benchmark process using dynamic principal components analysis based on decorrelated residuals (DPCA-DR).* Chemometrics and Intelligent Laboratory Systems, 2013. **125**(15): p. 101-108.

6. Arteaga, F. and A. Ferrer, *Dealing with missing data in MSPC: several methods, different interpretations, some examples.* Journal of Chemometrics, 2002. **16**: p. 408-418.

401632 Online Soft Sensor Adaptation of Latent Projection Growing Structure Multiple Model Systems

Abstract:

Online Soft Sensor Adaptation of Latent Projection Growing Structure Multiple Model Systems

Bo Lu[a], John Stuber[b], Thomas F. Edgar[a]

[a] McKetta Department of Chemical Engineering, [b] Texas Instruments Inc.

The University of Texas at Austin, 1 University Station C0400, Austin, TX 78712
Email: edgar@che.utexas.edu

Traditional semiconductor fabrication relies on run-to-run control with external metrology measurements for controller feedback. Virtual metrology is analogous to soft sensor, which aims to predict end-of-batch quality measurements using trace data and other end-point readings from semiconductor unit operations. Based on the predictions, control engineers can then make the decision on whether to perform additional metrology sampling or apply feed-forward adjustment in subsequent processing steps (Lu et al. 2014). Divide-and-conquer based data-driven modeling have been used in virtual metrology to better approximate nonlinearities, multiple steady states, and state transitions in the underlying process. Growing structure multiple model systems (GSMMS) uses a growing self-organizing map (GSOM) to divide the input space into discrete subspaces that can then be modeled locally (Liu et al. 2009). Previous work focused on extending GSMMS to use latent projection based models using PLS or PCA. The advantages of GSMMS over alternatives such as regression trees, multimode PLS (Dunia et al. 2012) or Gaussian Mixture (Yu and Qin 2008) based approaches are the lower degrees of freedom and more efficient use of training data. Since GSMMS adapts the complexity of the GSOM during training, there are less parameters to be specified. In addition, local model training uses data from an entire "neighborhood" to improve smoothness across local model boundaries.

In this work, an online adaptive update scheme for the latent projection GSMMS has been developed. Unlike previous multiple model systems, where the online adaptation mostly focus on local model coefficient updates, the proposed update algorithm will adapt all components of the GSMMS system to potential process changes. First, the local model coefficients for each node are updated recursively. Second, the growing self-organizing map nodes are shifted to adapt changes in the clustering layer. Lastly, additional nodes can be appended to the existing model system if a previously unseen process condition occurs at run-time. The proposed online adaptation mechanism has been tested on a simulated dataset and an industrial dataset. Using the industrial dataset, potential savings and reduction in necessary measurements were calculated to demonstrate the tangible economic benefits of the proposed approach.

References

Dunia, Ricardo, Thomas F. Edgar, Terry Blevins, and Willy Wojsznis. 2012. "Multistate Analytics for Continuous Processes." *Journal of Process Control* 22: 1445–56.

Liu, Jianbo, Dragan Djurdjanovic, Kenneth a. Marko, and Jun Ni. 2009. "A Divide and Conquer Approach to Anomaly Detection, Localization and Diagnosis." *Mechanical Systems and Signal Processing* 23(8): 2488–99. http://linkinghub.elsevier.com/retrieve/pii/S0888327009001873 (March 8, 2012).

Lu, Bo, John Stuber, and Thomas F. Edgar. 2014. "Integrated Online Virtual Metrology and Fault Detection in Plasma Etch Tools." *Industrial & Engineering Chemistry Research* 53(13): 5172–81. http://pubs.acs.org/doi/abs/10.1021/ie401448d.

Yu, Jie, and S Joe Qin. 2008. "Multimode Process Monitoring with Bayesian Inference-Based Finite Gaussian Mixture Models." *AIChE Journal* 54(7): 1811–29.

401634 Visualization and Fault Detection in Radial Coordinates

Abstract:

Visualization and Fault Detection in Radial Coordinates
Ray Wang, Michael Baldea, and Thomas F. Edgar

McKetta Department of Chemical Engineering

The University of Texas at Austin, 1 University Station C0400, Austin, TX 78712

Mark Nixon, Willy Wojsznis, Ricardo Dunia

Emerson Process Management, Austin, TX

Email: mbaldea@che.utexas.edu

With the advancement of technology, many industries are devising new methods to produce and meet demand more efficiently. Data on various processes, be they chemical, manufacturing, or pharmaceutical, are collected using a variety of sensors. As processes increase in complexity to meet demand for innovative products, so does the number of sensors needed to keep track of the process. Furthermore, more precise control of the process may be needed, which leads to an increase in the sampling frequency of the sensors. Both of these trends lead to an increase in the volume of data collected.

As a consequence, process operating data have entered the "big data" age. While these data are readily available in databases or process historians, extract useful information for process analysis and decision support remains a difficult task. An intuitive first step to understanding data is visualization; this is, however, challenging to accomplish for big process data using the customary 2D score plots.

To answer this challenge, in this paper we propose a variation of the parallel coordinate system [1], which we will refer to as 3D radial plots. 3D radial plots are based on using a closed-polygonal line to represent each data sample (as opposed to an open line used in parallel coordinates). The polygons representing data samples are "stacked" vertically according to the time data were acquired, which provides an explicit representation of time. Moreover, this representation allows for plotting of an arbitrary number of time-dependent variables on a single plot.

We exploit the geometric properties of the 3D radial plot framework and discuss several extensions, related to performing tasks such as fault detection and monitoring the evolution of a process during significant transitions between operating points. Further, we establish a connection between the proposed fault detection methodology and multivariate control charts.

The theoretical concepts are illustrated using industrial examples, amongst which, monitoring flooding in a distillation column and detecting surge events in a compressor.

References:

[1] A. Inselberg, The plane with parallel coordinates, Vis. Comput. 1 (1985) 69–91.

401958 Integrated Grid Modeling System: A Distributed Solar Power Case Study

Abstract:

The future electricity system is a distributed control problem on an enormous scale: millions of generation and load devices requiring coordinated but decentralized decision making. The interactions between a residential rooftop PV system and the household appliances will feed up into the local distribution system. The interactions among the different residential and commercial systems feed into a distribution level supply or demand that will be further fed up into the transmission system. At the transmission level coordination is necessary to be able to provide ancillary services or to link sub-systems where supply and demand are mismatched with those that currently have the opposite power balance due to fluctuating conditions. The cyber (information flow) layer architecture formalizing the interactions between these different systems needs to be developed while respecting that the physical electricity interactions are dictated by the physics of the electricity and the communication systems.

Historically in the transmission and distribution networks have been modeled independently of one another due to computational limitations. However, this does not allow for an understanding of the impact of distributed devices on bulk system operations, nor of the impact of transmission level policies on distributed generation. In this work we have developed a simulation platform that ranges from the appliance-level to the Independent System Operator (ISO) level which allows for the investigation of these impacts. The open source software GridLab-D is utilized for distribution level system modeling, and has been linked with the FESTIV transmission level model to represent both ends of the power system operations spectrum. As most transmission level substations have multiple distribution feeders, and the typical timescale for feeders is much lower than for bulk system operations, thousands of GridLab-D instances must be run to supply the information for each FESTIV time step. These simulations are only possible to be completed in realistic timescales due to the use of NREL's Peregrine high performance computing system.

The testbed has been utilized to study the impacts of high penetration distributed solar power on transmission level system operations. Distributed solar power installations are currently growing very quickly in a number of areas of the United States, and are beginning to impact transmission system operations. For example, while system operators generally do not have any visibility into distributed solar installations or their power output, and only have visibility into their output based on deviations from their load forecasts. The case study examined includes the impact of high penetration distributed solar power on the unit commitment and economic dispatch process, bus voltage levels, and distributed solar power forecasting, among other options.

401959 Effect of Cryogenic Carbon Capture (CCC) on Smart Power Grids

Abstract:

Increased electricity demand and stability concerns for existing grids have led researchers to seek for power generation systems that can relieve the pressure from the existing grid while also satisfying environmental concerns. The smart grid is considered to be a new strategy to achieve this goal [1]. While there is not a universal definition for smart grids, it usually includes data collectors, data displayers, information-based controllers, and technologies that can produce and store electricity or reduce the demand [2]. More specifically, smart grids consist of advanced metering infrastructures [3], advanced electricity pricing [4], demand response [5], distribution automation [2], microgrids [6], and distributed storage systems [7]. These technologies stabilize power grids and enable them to accommodate more renewable power sources [2]. Smart grid technologies attract more attention when CO2 emission regulations become stricter because CO2 removal systems consume a significant fraction of the power plant output. While management of consumer demand is often considered in smart grid studies, this strategy may be applied once or twice a year for a few hours where there is peak electricity demand [8]. Supplier load management applies the same idea and has the potential to be used throughout the year, especially when the continuous operation of CO2 removal systems is considered. This strategy can therefore deliver more power to the grid during peak hours and stabilize it without overriding consumer choice.

Cryogenic Carbon Capture TM (CCC) is an example of a CO2 removal system that stabilizes the grid by storing available excess energy and using it during peak hours [9]. This process primarily includes fast-responding equipment such that it can follow the high fluctuations of the intermittent renewable power sources. This energy storage incorporates renewable sources into the grid by storing them when they are available [10]. Therefore, the excess energy can originate either from baseline power production or from renewable power sources. In this process, the electricity required to run the CO_2 removal plant during peak hours is supplied from the stored energy during off-peak hours. The stored energy can also be used in meeting the residential demand. Therefore, the energy storage capability of the CCC process will not only relieve excess stress on the grid during peak hours but also relieves the pressure from it by delivering the stored power to the grid. The importance of such multi-functional plants significantly increases when the restrictive regulations for CO_2 emissions become effective. Thus, technologies such as CCC require detailed attention.

In the CCC process, CO_2 is removed from the flue gas by desublimation. To provide the cooling for the desublimation process, two refrigerants are used in the CCC process. Most of the energy consumption of the CCC process is attributed to the compression cycles. However, this process has the potential to produce excess refrigerants when electricity demand is low or when excess energy is available and store them in grid-scale insulated vessels. Liquefied Natural Gas (LNG) is the refrigerant used in this study to run the CCC process in energy storage/recovery modes. The storage/recovery modes of the second refrigeration cycle is not considered. LNG is produced in excess during low electricity requirement and stored in the tank. Then, during peak hours, stored LNG is taken from the tank to run the CCC process. Running the CCC process with stored LNG means that refrigeration compressors are run at the minimum load which saves compressor power demand. After going through the CCC process, LNG becomes a vapor as a result of removing heat from the flue gas. The excess natural gas produced inside the plant during peak hours is partially combusted in a gas turbine to provide electricity for both the CCC process and residential consumers. The rest of the natural gas can be either exported to the pipeline to avoid processing it during peak hours, further saving electricity for the grid, or can be recirculated to the LNG production plant to supply a fraction of the LNG for the CCC process. Therefore, the storage capability of the CCC process helps stabilize the grid during peak hours by providing excess power and reducing the parasitic loss of the CO_2 removal system.

In this case, dynamic optimization and integration of the CCC process with fossil-fueled power plants and wind power plants is considered. The coal-fired power plant has a 900 MW capacity while the power output from the gas turbine is 480 MW. The wind power comes from ten stations in California and reaches a maximum of 300 MW. Dynamic optimization results show that the assumed grid-scale energy storage tank manages the electricity load of the CCC process such that approximately 80 MW excess power for about 12 hours of the day is available for the grid during peak hours. In addition, approximately 570 MW power is also produced from the natural gas produced inside the plant (gas turbine and combined cycle), again for about 12 hours. The operational profit obtained from the system increases by 122% compared to the same system without energy storage. The excess revenue obtained from the storage tank is enough to pay for the cost of construction of the CCC plant [11].

[1] "Electrical Engineernig Portal." Available: http://electrical-engineering-portal.com/an-overview-of-smart-power-grid.

[2] "Smart Grids and Renewables; A Guide for Effective Deployment", International Renewable Energy Agency (IRENA), 2013.

[3] H. Farhangi, "The Path of the Smart Grid", Power and Energy Magazine, IEEE , vol.8, no.1, pp.18,28, Jan.-Feb. 2010.

[4] S. Shao, M. Pipattanasomporn, "Impact of TOU Rates on Distribution Load Shapes in a Smart Grid with PHEV Penetration", Transmission and Distribution Conference and Exposition, 2010 IEEE PES , vol. 1, no. 6, pp. 19-22, Apr. 2010.

[5] A. Mahmood, M. N. Ullah, S. Razzaq, and E. Al., "A New Scheme for Demand Side Management in Future Smart Grid Networks", Procedia Computer Science, vol. 32, pp. 477–484, 2014.

[6] H. Kanchev, D. Lu, F. Colas, and B. Francois, "Energy Management and Operational Planning of a Microgrid With a PV-Based Active Generator for Smart Grid Applications", Industrial Electronics, IEEE Transactions, vol.58, no.10, pp.4583,4592, Oct. 2011.

[7] P. M. Connor, P. E. Baker, D. Xenias, et. al., "Policy and regulation for smart grids in the United Kingdom", Renewable and Sustainable Energy Reviews, vol. 40, pp. 269–286, Dec. 2014.

[8] "Enernoc", Available: http://www.enernoc.com/our-resources/term-pages/what-is-demand-side-management.

[9] "Sustainable Energy Solutions", Available: http://sesinnovation.com/.

[10] M. J. Jensen, D. Bergeson, D. Frankman, L.L. Baxter, "Integrated Rapid Response Energy Storage with CO2 Removal", PowerGEN, 2012.

[11] S. M. Safdarnejad, J. D. Hedengrena, L. L. Baxter, "Grid-level Dynamic Optimization of Cryogenic Carbon Capture with Conventional and Renewable Power Sources", Applied Energy Journal, submitted.

402078 Energy Demand Response of Process Systems through Production Scheduling and Control

Abstract:

As the availability and pricing of electricity vary during the day due to intermittency in renewable resources or real-time electricity tariff policies, the chemical industry, especially those with power-intensive processes, needs to rethink operating strategies to maximize economic viability. The implementation of demand response (DR) strategies, which focuses on demand side activities and has been in the center of all pragmatic energy policy decisions, is the key to minimizing production costs and maximizing profits. However, the dynamic behavior, especially transition trajectories, of the underlying process is seldom taken into account during this task. The purpose of this study is to present a novel optimization formulation for energy demand management in dynamic process systems that takes transition behavior and cost explicitly into account, while simultaneously handling time-sensitive electricity prices. This is accomplished by bringing together production scheduling and transition control through a real-time optimization framework. The dynamic formulation is cast as a mixed-integer nonlinear programming problem and demonstrated using a chemical reactor example where the energy required is assumed to be roughly proportional to the material flow.

399667 Manufacturing Analytics – Conquering "Big Data" in Chemical Manufacturing

Abstract:

The Dow Chemical Company has been on a developmental path for three years now to fundamentally change how we handle and analyze large sets of real-time data to improve plant reliability and lower manufacturing costs. Introducing a concept at ARC - Orlando, 2013, we presented a talk on the design of an Enterprise Manufacturing Intelligence within the Dow Chemical Company[i]. This talk focused the burning platform behind the initiative, visioning process that led to the initial steps of the implementation of an Enterprise Manufacturing Intelligence (EMI) platform within Dow, and described some initial challenges, and some preliminary results of this vision.

Since this time, Dow has continued the journey in the implementation of an EMI platform and is achieving success at implementing an enterprise platform that is bringing recognized value to our businesses and their manufacturing assets.

In this paper, we will review the fundamental approaches that needed to change to bring about the transformation we have achieved in mastering our plant data, provide use cases for the implementation of our enterprise design, as well as provide examples of potential future analytics cases as we attempt to continue the push for smarter, greater value analytics approaches to our plant data. We will also provide some insight on what the future may hold as we drive towards a broad process to convert data to information to knowledge to wisdom.

---

[i] Dow Chemical Uses Analytics and EMI to Transform Big Data into Knowledge, J. Abel, ARC Insights, June 2013

Business Insight from Big Data in the Chemical Industry

Erika McBride
Dow Chemical Company

Justin Kauhl
Tata Consultancy Service

15 AIChE
Spring Meeting, Austin, TX

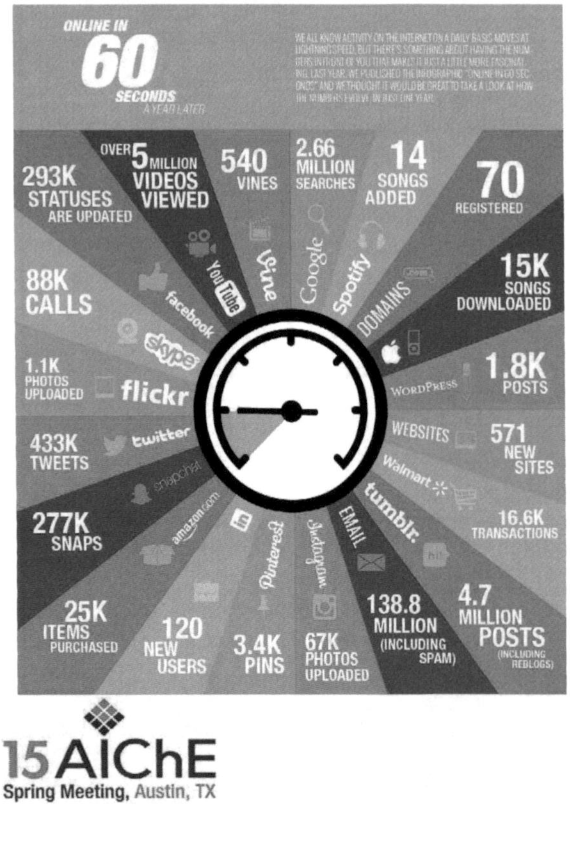

Qmee

DATA
www.internetlivestats.com
www.thesocialskinny.com
www.econsultancy.com
www.cnn.com
www.wikibon.org
www.flickr.com
www.weidert.com

www.linkedin.com
www.tumblr.com
www.quora.com
www.amazon.com
www.iacpsocialmedia.org
www.flickr.com
www.mashable.com

DESIGN BY NoLimitAgency

15 AIChE
Spring Meeting, Austin, TX

Advanced Analytics

Translating Data Into Profit

Forecasting	Optimization and Simulation	Data and Text Mining
Plan for the future.	Achieve the best outcomes.	Find meaningful patterns and relationships in structured and unstructured data.

15 AIChE Spring Meeting, Austin, TX

DOW RESTRICTED - For internal use only

Courtesy : SAS for the orange icons

Social Media Mining

15 AIChE Spring Meeting, Austin, TX

Social Media Mining

- How can we use **social media data** to gain insight into consumer markets?
- Can we use this insight to make better products?

Brand Reputation

- What do consumers think about brand A compared to brand B?
- What is the aspect of a brand that consumers like the most?

Product Aspects

- What aspects of a product do customers mention when choosing a brand?
- Can social media tell us what aspect of a product matters the most?

Consumer Sentiment

- What do consumers care most deeply about?
- What are their biggest complaints?

15 AIChE
Spring Meeting, Austin, TX

DOW®

Social Media Mining

- **Public Product Pages and mentions**

Identified Categories of Message Topics

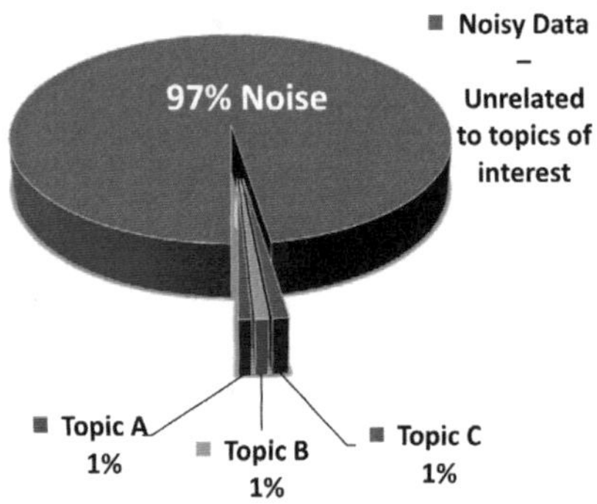

<u>Very low signal to noise ratio!</u>

15 AIChE
Spring Meeting, Austin, TX

DOW®

Social Media Mining

- **Remaining posts with topics of interest present**

"Which brand has the best reputation for the target topic? Why?"

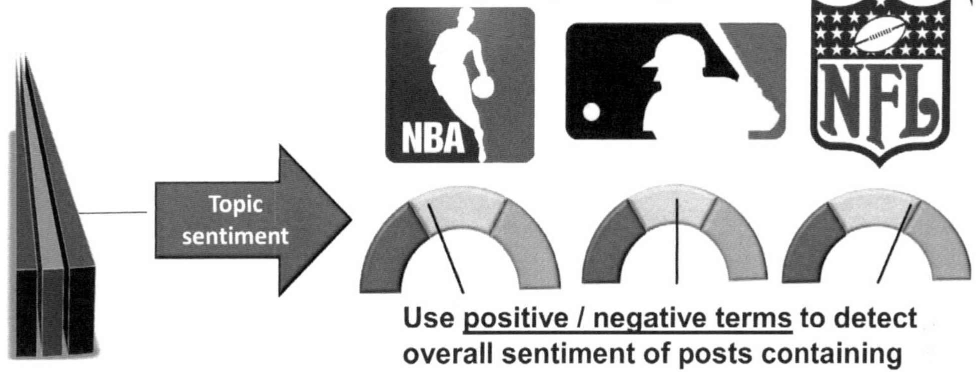

Use positive / negative terms to detect overall sentiment of posts containing identified topic by brand

15 AIChE
Spring Meeting, Austin, TX

DOW®

Social Media Mining

Which NFL are you interested in?

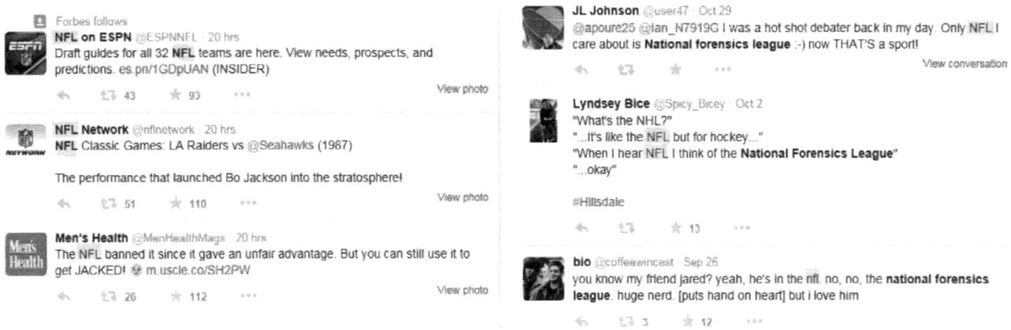

Forbes follows
NFL on ESPN @ESPNNFL · 20 hrs
Draft guides for all 32 **NFL** teams are here. View needs, prospects, and predictions. es.pn/1GDpUAN (INSIDER)
43 93
View photo

NFL Network @nflnetwork · 20 hrs
NFL Classic Games: LA Raiders vs @Seahawks (1987)

The performance that launched Bo Jackson into the stratosphere!
51 110
View photo

Men's Health @MenHealthMags · 20 hrs
The NFL banned it since it gave an unfair advantage. But you can still use it to get JACKED! m.uscle.co/SH2PW
26 112
View photo

JL Johnson @user47 · Oct 29
@apoure25 @Ian_N7919G I was a hot shot debater back in my day. Only NFL I care about is **National forensics league** :-) now THAT'S a sport!
View conversation

Lyndsey Bice @Spicy_Bicey · Oct 2
"What's the NHL?"
"...It's like the NFL but for hockey..."
"When I hear NFL I think of the **National Forensics League**"
"...okay"

#Hillsdale
13

bio @coffeewincest · Sep 26
you know my friend jared? yeah, he's in the nfl. no, no, the **national forensics league**. huge nerd. [puts hand on heart] but i love him
3 12

15 AIChE
Spring Meeting, Austin, TX

DOW®

Social Media Mining

- Capable of zeroing in on one particular topic
- Able to predict what our customers need by mining their customer's needs
- Rapid reaction to changes in market sentiment
  - No need to wait for problems to ripple up the value chain!

15 AIChE
Spring Meeting, Austin, TX

DOW®

Economic Sentiment

15 AIChE
Spring Meeting, Austin, TX

DOW®

Economic Sentiment

- Can we use financial report topic sentiment to **improve models**?:
  - Minimize Latency
  - Increase Quality
- Utilize unstructured data sources
  - Repository of thousands of Report PDFs
  - Web-crawl newsfeeds for current data

15 AIChE
Spring Meeting, Austin, TX

DOW®

Economic Sentiment

- Work with SMEs to identify **sources** and **specific topics** that can improve forecast accuracy or extract information

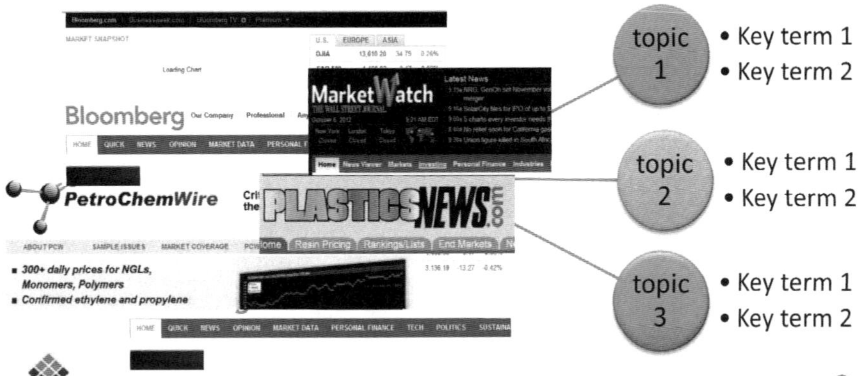

15 AIChE
Spring Meeting, Austin, TX

DOW®

Economic Sentiment

- At sentence level, sentiment, geography, and concept topic weights are generated

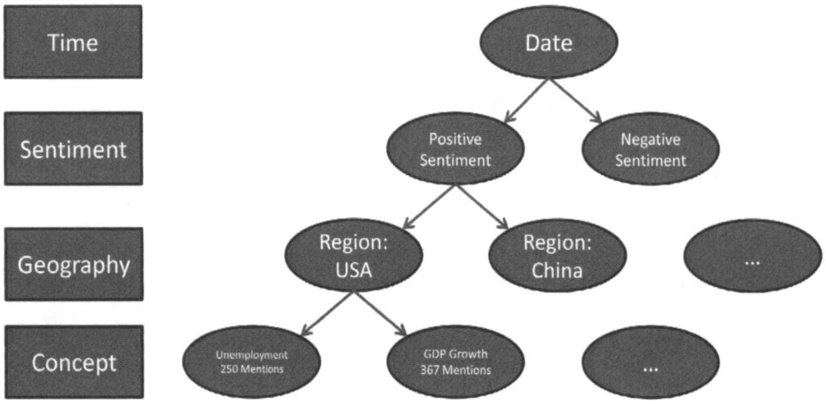

15 AIChE Spring Meeting, Austin, TX

DOW®

Economic Sentiment

- Positive / Negative Aggregate Weekly Sentiment for China Industrial Production

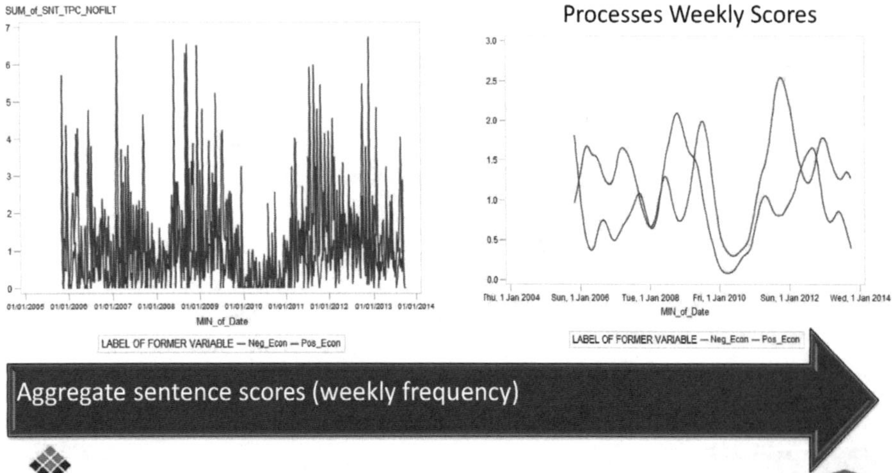

15 AIChE Spring Meeting, Austin, TX

DOW®

Economic Sentiment

- **Aggregated Weekly Sentiment for China Industrial Production**

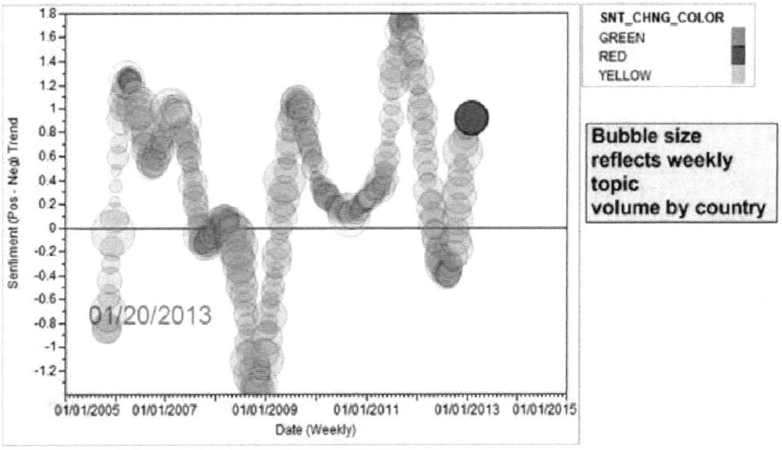

15 AIChE
Spring Meeting, Austin, TX

DOW

Economic Sentiment

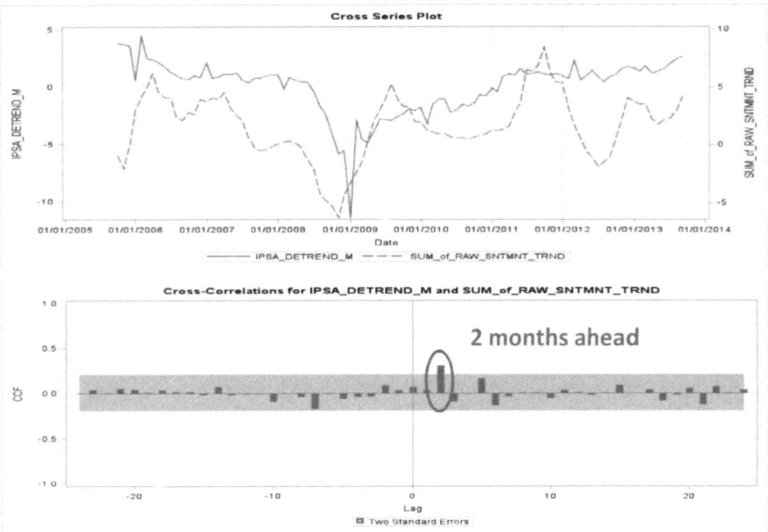

15 AIChE
Spring Meeting, Austin, TX

DOW

Economic Sentiment

- News feed economic sentiment is a quality leading indicator
- ~5% reduction in model error
- Can model at a higher frequency
  - Lower lag times

15 AIChE Spring Meeting, Austin, TX

DOW

Questions?

15 AIChE Spring Meeting, Austin, TX

DOW

Thank You!

Erika McBride
Email : EMcBride@dow.com
Twitter : @McBride_Erika

Justin Kauhl
Email : Justin.Kauhl@tcs.com

15 AIChE
Spring Meeting, Austin, TX

400388 Learning from Historical Data: A Key to Process Improvement and Optimization

Abstract:

Process and laboratory computers collect large amounts of data. Knowing how to extract actionable information from these data is crucial. This seminar will look at recent advances and applications in the industrial use of multivariate data analysis methods for process improvement and optimization.

A brief conceptual introduction will be given to the latent variable / chemometric modeling methods used and why they are so much more suitable for the analysis of historical process data than traditional statistical methods.

Industrial examples will include the identification of important process and raw material property effects on final quality in a continuous process, and troubleshooting and optimizing a batch specialty chemical process.

401960 Predictive Strategies for Coordinating Passive and Active Thermal Energy Storage in Buildings

Abstract:

Residential and Commercial buildings account for over 70% of the electricity demand in the US [1], and a significant portion of this is used towards maintaining indoor comfort via heating, ventilation, and air-conditioning (HVAC). Building electricity use fluctuates during the day and exhibits a peak in the afternoon. This peak is reflected in a demand maximum at the grid level, which is in general costly and inefficient for grid operators and energy producers to satisfy [2]. Ideally, this variation in energy demand on the grid should be addressed using energy storage. Buildings constitute an ideal candidate to this end, owing to their inherent capability to store thermal energy in their structural elements. This feature, which we refer to as "passive" thermal energy storage can be augmented using active TES systems consisting, e.g., of water or ice tanks [3].

TES systems can be charged at off-peak hours, when energy costs are low [4], and the stored energy can be released/discharged during peak hours. In this presentation, we focus on devising optimal strategies for coordinating charging and discharging events with the operation of the grid, while, on the other hand, meeting comfort constraints imposed by the building occupants. The dynamics of the storage systems are typically slow and are dictated by the thermal inertia of the building and, respectively, by the dynamics of the active storage capacity. On the other hand, the building is subject to relatively fast disturbances from changes in occupancy, insolation, and ambient temperature. Thus, coordinating the operation of buildings with TES is a multi-scale problem. Additional complications arise from the complexity of the dynamic models of each subsystem (building, storage tank, etc.) and from the need to account for discrete decisions related to the operating mode (charge, discharge) of the TES.

In this work, we propose a novel approach, where we decompose the optimal operation problem into (i) a scheduling problem for the active TES in a slow time scale, and (ii) a MPC with a shorter horizon in the fast time scale to address the use of passive TES [5]. We show that this proactive decision making framework is superior to heuristics for controlling the TES in the smart grid environment. Moreover, a Monte Carlo simulation demonstrates that, when implemented across many buildings, the proposed approach has a leveling effect on grid demand from HVAC systems. In a broader context, this leads to a reduction in the environmental impact of buildings by minimizing the on-time of inefficient "peaking plants" that are used to satisfy peak demand.

[1] U.S. Energy Information Administration, Annual Energy Outlook 2013

[2] J. Siirola, T. Edgar, Process energy systems: control, economic, and sustainability objectives, Comput. Chem. Eng. 47 (2012) 134–144.

[3] G. P. Henze, C. Felsmann, G. Knabe, Evaluation of optimal control for active and passive building thermal storage, Int. J. Therm. Sci. 43 (2) (2004) 173–183.

[4] M. H. Albadi and El-Saadany E. F. Demand response in electricity markets: An overview. Proc. IEEE Power Eng. Soc. Gen. Meet., pages 1–5, 2007.

[5] C. Touretzky, M. Baldea, Integrating scheduling and control for economic MPC of buildings with energy storage, J. Proc. Contr. , 24 (8) (2014) 1292-1300.

402077 Industrial Demand Side Management Under Uncertainty

Abstract:

With deregulated electricity markets and increasing penetration of intermittent renewable energy into the electricity supply mix, the level of uncertainty in the power grid has increased tremendously. This has led to highly volatile electricity prices, which pose immense challenges to the power-intensive industries, such as air separation, aluminum, and chlor-alkali manufacturing. Demand Side Management (DSM), which refers to electric energy management on the consumers' side, has the potential of both significantly reducing the electricity cost as well as improving the efficiency and reliability of the power grid for the consumer.

The high potential impact of large industrial electricity consumers participating in DSM is widely acknowledged (Samad & Kiliccote, 2012; Merkert, et al., 2014) and has been the focus of increased research efforts in recent years. Undoubtedly, one of the major challenges in DSM is uncertainty, which does not only appear in electricity prices, but also in power availability, product demand, and the occurrence of demand response events etc. Robust and economically sound decisions can only be made when these uncertainties are appropriately taken into account in the decision-making process.

In this work, we develop optimization tools for two industrial DSM applications in which modeling uncertainty is crucial. In the first application, we consider the integrated optimization of production scheduling and electricity procurement, in which we determine how much electricity to purchase from signed bilateral contracts and how much from the spot market. Here, we apply stochastic programming (Birge & Louveaux, 2011) to account for uncertainty in spot electricity price and product demand. In the second application, we consider an air separation plant that utilizes cryogenic energy storage to store electricity and recover it later for internal use or to sell it back to the grid. By appropriately modeling the uncertainty in operating reserve demand using robust optimization techniques (Ben-Tal, et al., 2009), we can also use the energy storage capability to sell reserve capacity, which can be requested by the grid operator in times of emergency. In both cases, we apply the proposed approaches to real-world industrial case studies and demonstrate the value of accounting for uncertainty.

References

Ben-Tal, A., El Ghaoui, L. & Nemirovski, A., 2009. *Robust Optimization.* New Jersey: Princeton University Press.

Birge, J. R. & Louveaux, F., 2011. *Introduction to Stochastic Programming.* 2nd ed. New York: Springer Science+Business Media.

Merkert, L. et al., 2014. Scheduling and energy - Industrial challenges and opportunities. *Computers & Chemical Engineering,* Volume 72, pp. 183-198.

Samad, T. & Kiliccote, S., 2012. Smart grid technologies and applications for the industrial sector. *Computers & Chemical Engineering,* Volume 47, pp. 76-84.

402075 On the Use of Multistage Stochastic Programming for the Design of Smart Grid Coordinated Systems

Abstract:

Due to the uncertainty of electricity prices, the problem of optimal equipment sizing for smart grid coordinated process must be formulated as a stochastic program. However, rather than being a fairly simple two-stage stochastic program, the dynamics imposed by the storage devices requires the formulation to be of the far more challenging multistage class.

In this work, we propose a novel approximate solution procedure for this class of multistage stochastic programming problems. The approach utilizes the computational efficiency of the recently developed method of Economic Linear Optimal Control (ELOC) and its extension Constrained ELOC. The proposed method occurs in two stages. The first is a global search over the here-and-now variables as well as the parameters of the ELOC policy. However, this first search assumes a statistical constraint enforcement mechanism (similar to chance constrained optimization). In the second step is a gradient based search over the here-and-now variables, but used the Constrained ELOC policy to enforce point-wise-in-time constraints. The proposed method will be illustrated using the example of an IGCC plant that has be made dispatch capable through the installation of gas storage units.

401334 Predictive Asset Optimization and Intelligent Decision Making Utilizing Big Data to Enable Smarter Manufacturing

Abstract:

The concept of Big Data is not new to oil & gas, downstream and petrochemical industries. In any operating facility there are thousands of sensors that are measuring process data every few seconds. This huge volume of data that is collected is typically stored in information management systems. In addition, large amounts of events data such as alarms, alerts and plant maintenance & reliability information such as issues, notifications, reports, work orders, failures are stored in the alarm / alert management systems and maintenance management systems.

The oil & gas, downstream and petrochemicals industries are facing challenges such as aging workforce with wide gaps in experience, aging asset base with more complex operations, market variability and speed of change requiring faster and better decisions at real-time, increased competition driving operational excellence, production and energy optimization and sustainability, increasing environmental regulations requiring effective controls, safe, secure and reliable operations for compliance.

The limited resources available are more focused on day-to-day routine operations, maintenance and response. There are no spare resources available to invest the time to analyze the vast amount of existing sensor and event data / information to proactively identify / discover trends and patterns and utilize those insights to optimize manufacturing operations and to improve the reliability of critical assets.

Given these challenges and the context of the emerging convergence of Information Technology and Operations Technologies, we will discuss using various example use cases from oil & gas, downstream and petrochemicals industries, how emerging technologies such as Predictive Analytics and Cognitive Computing (e.g. Watson) utilize big data to improve performance, profitability, efficiency, quality, safety and asset availability. Utilizing these technologies, we can analyze and correlate a combination of all available data and information in various formats to identify trends and patterns and / discover insights. Based on this improved predictive operational intelligence visualized through advanced operations and maintenance dashboards, the operations and maintenance resources can be more effective and enabled to make the right decisions in real-time to increase the availability and utilization of critical equipment leading to optimized production, reduced unplanned capacity loss, reduced quality impacts, improved safety and reduced overall maintenance cost.

EMBEDDED ANALYTICS IN INDUSTRIAL BIG DATA APPLICATIONS

John Caldwell, Mark Nixon, Terry Blevins and Willy Wojsznis
Innovation Center, Emerson Process Management, Austin, TX

Extended Abstract

The chemical process industry is challenged by increasing technical complexity involving large integrated facilities and rapidly evolving process automation technologies. Experienced workers are needed to make risk decisions yet much of the experienced personnel are nearing retirement and new plants are often located in areas which lack an experienced workforce. Process analytics is a powerful tool to gain better insight into these complex operations, to improve performance and reliability, and enhance operator effectiveness. However, practical implementation of on-line analytics can be problematic because of real-time data integration, operations work flows, and difficulty in implementation and maintenance of traditionally layered solutions.

This presentation describes an approach for embedding data analytics in a control system to provide on-line process analytics for improved operator effectiveness, production performance, and manufacturing reliability. The embedded solution is configured and implemented in the native control system to reduce complexity and provide a consistent operations interface. Seamless integration with real-time and historical data simplifies model training and on-line implementation using standard control system engineering tools. The presentation includes results from field trials in polymer production and oil refining facilities.

Capabilities of the embedded analytics include process quality parameter prediction and abnormal situation monitoring with fault detection. Statistical modeling technologies used for process parameter prediction include Projection to Latent Structures (PLS), Neural Networks (NN), and Multiple Linear Regression (MLR). Abnormal situation and fault detection incorporates Principal Components Analysis with enhancements for multi-state operations. The embedded analytics are executed as function blocks inside the control system controllers and are configured and trained using techniques familiar to control system users.

Embedded data analytics is a core component of big data applications in the process industries and can be used for targeted data analytic applications which benefit from residing in the control system where they can be easily integrated into other control system applications and operations interfaces. An integrated approach toward on-line analytics provides an easy-to-use set of tools for analytic model development and verification, with subsequent download to the DCS controller for on-line operation. The model development and download for on-line operation is streamlined into clearly defined and easily executed steps:

- Defining analytic model configuration, i.e. identifying potential process parameters for PCA/PLS model
- Creating analytic module based on the defined model configuration, downloading module to the DCS controller.
- The module is configured to collect history data, including lab analysis that is defined by the analytic model.
- Developing analytic models, including PCA, PLS, NN and MLR from collected historical data as defined by the analytic module or from an external historical data file created prior to the analytic module download

- Validating and downloading model for on-line operation

On-line analytics monitors faults in the process operations and predicts product quality. The results are presented in a web based interface and can also be used to enhance an existing alarming system. The predicted property quality can be used in advanced control and optimization strategies such as model predictive control.

The continuous data analytic design has been tested on simulated data and in industrial on-line applications in the field. In one such field application, an embedded analytics PLS model was used to predict heavy components in a distillation column. When the results were compared with an online analyzer, the analytic modeling demonstrated several advantages: more reliable operation, less demanding maintenance, and integrated fault monitoring which supports quality prediction validation and process operations troubleshooting. Process engineers greatly appreciated the ability to identify abnormal distillation column operation and the potential for including this functionality into an existing alarming system.

The field trial facilitated the development of an iterative procedure for analytic model improvement. Similar results were obtained from a polybutene unit analytic model development and an on-line test. Specific to this process is the product quality (viscosity) which can be set on two distinct levels, depending on the manufacturing needs. Two alternative approaches were explored: one using a state parameter associated with viscosity and the other using two models for Low and High viscosity values. In the majority of tests two separate analytic models performed better. The primary problem with the multistate analytic model was identifying a process parameter that was defined in the analytic model as a state parameter that correlates well with quality level.

Acknowledgment

We greatly appreciate the professional and extremely dedicated support in the field testing we received from Ashok Dasgupta and Tom Vernor from Huntsman and from Efren Hernandez and Bob Wojewodka from Lubrizol. James Beall from Emerson provided a wealth of control expertise and helped us to develop and install analytic models.

References

1. Terrence Blevins, Willy K. Wojsznis and Mark Nixon *Advanced Control Foundation,* ISA, 2013
2. Dunia, R., Edgar, T., Blevins, T., Wojsznis, W., *Multistate PLS for Continuous Process Monitoring,* ACC, March, 2012
3. Dunia, R., Edgar, T., Blevins, T., Wojsznis, W., *Multistate Analytics for Continuous Processes*, Journal of Process Control, 2012
4. MacGregor J.F., Kourti T., *Statistical process control of multivariate processes*. Control Engineering Practice 1995; 3:403-414
5. Kourti, T. *Application of latent variable methods to process control and multivariate statistical process control in industry*. International Journal of Adaptive Control and Signal Processing 2005; 19:213-246

400949 Leveraging Big Data with Self-Service Analytics / Tableau Software

Abstract:

You are an innovator. You work hard and want your organization to succeed but are short on time. For you and your organization, data-driven decision-making is more than just a buzzword — BIG data is increasingly critical to accomplishing your most important objectives and staying ahead of the competition. You understand the details and nuances of your marketplace but don't necessarily have the word "analyst" in your title. You use software like Microsoft Excel or business intelligence (BI) tools and have seen how much time it takes to pull in data from different sources, find insights, and share reports or dashboards. You believe there must a better way. This presentation is for you. It outlines a new approach to help you, your team and your organization see and understand BIG data faster and more easily.

402520 Integrated scheduling and dynamic optimization for semi-batch polymerization processes

Abstract:

In this work, we propose an integrated optimization approach for an industrial semi-batch polymerization process. The method introduces a discrete time formulation for the simultaneous optimization of process scheduling and operation decisions. The resource task network (RTN) representation is applied to model the interconnection between units and production sequences. Polymerization reactors are described by first-principles models addressing heat/mass balances, reaction kinetics, etc.

Our modeling framework accommodates general complications in scheduling and unit operation, such as customer orders, transfer policy and requirements on product quality and process safety. In the mathematical formulation, the scheduling and unit operation equations are linked with the task history state variables in the state space RTN model. A tailored generalized Benders decomposition (GBD) algorithm is applied to efficiently solve the resulting large nonconvex mixed-integer nonlinear program by exploring the particular model structure.

The studied polymerization process has two parallel semi-batch reactors for ring-opening polymerization, continuous storage tanks, and purification units. The two polymerization reactors share cooling utility from the same source, and the utility price depends on the consumption rate. The optimization objective is to design the process schedule and reactor control policies simultaneously to maximize the overall process profit. The case study results suggest improvements in plant profitability for the integrated approach, in contrast to the conventional approach, where recipes of the polymerization tasks are individually optimized and the interactions among process units are overlooked.

402522 Mixed-integer Optimization Formulations for Site-wide Turnaround Planning

Abstract:

Planned chemical plant turnarounds are essential activities that ensure continuing reliability and productivity. However, turnarounds represent production disruptions for significant periods of time and, in a large chemical site network, turnarounds also affect site integration. It is of utmost importance to consider the planning and scheduling of the turnarounds to minimize these effects through appropriate manpower allocation, production, and inventory decisions. We present a mixed-integer linear programming framework for turnaround planning with long, medium and short-term considerations. We include site network structure, turnaround decisions, and several practical constraints of industrial relevance, and provide industrial case studies.

400421 Using Production and Image Data for Real-Time Improvement of Process Operations

Abstract:

DCS and PLC systems collect large amounts of data routinely every few seconds. More complex PAT and image based sensors are also in more common use. Knowing what these data are telling you about your process so that the process operations can be improved, and so that operators or control systems can respond in real-time is critical to improved plant performance. This presentation looks at recent advances and applications in the industrial use of on-line multivariate data analysis systems for process improvement, monitoring and control. These include:

(1) Real-time monitoring of batch specialty chemical processes.

(2) Predictive control of final product quality in batch processes using multivariate data based models and mid-course corrections.

(3) Real-time inspection of multiple defects on synthetic rubber bales via advanced multivariate image analysis and relating the defect severity results to upstream processing conditions to significantly improve the process.

399936 Moving from Inspect and Reject to Predict and Prevent: Incorporating Real-Time Analytics into Your Manufacturing Alarm Management Strategy

Abstract:

As we enter the world of Big Data most manufacturers are drowning in a sea of too many alarms. This presentation will discuss how a growing number of companies are leveraging real-time manufacturing analytics to improve product quality and plant efficiency. We will highlight several approaches, and applications for determining appropriate alarm types and how they are being used to produce a powerful return on investment. Applying the ANSI/ISA 18.2 Alarm Management Standard, this presentation will provide best practices for work processes, which provide a path for continuous improvement, while leveraging AC/CA (CAPA) tracking. We will explore prioritization, acknowledgment, SMS/email notification, and advanced alarming capabilities. Finally we will address the right implementation practices to develop organizational habits which maximize the effectiveness of any system.

Jim Petrusich, NWA Software

399462 Alamo: Automatic Learning of Algebraic Models Using Optimization

Abstract:

We address the problem of discovering algebraic relationships that are hidden in a set of data, an experimental process, or a simulation model. The problem lies at the interfaces between statistical experimental design, optimization, and machine learning. We present a methodology for developing models that are simple and accurate, while minimizing the number of experiments or simulations of the system under study. The methodology begins by building a low-complexity model of the system using integer optimization techniques. The model is then tested, exploited, and improved through the use of derivative-free optimization to adaptively sample new experimental or simulation points. Semi-infinite optimization techniques facilitate a combined data- and theory-driven approach to model building. We provide computational comparisons between ALAMO, the computational implementation of the proposed methodology, and a variety of machine learning and statistical techniques, including Latin hypercube sampling, simple least squares regression, and the lasso. Finally, we demonstrate how ALAMO's adaptive sampling technique can be used to learn models by selecting small numbers of samples from huge data sets or even from infinitely many data points.

402525 Using modeling to control a chemical plant flare

Abstract:

In a chemical plant, a flare is often used to dispose of production waste by-products. The heating value of the waste products is often not high enough for them to burn at the target combustion temperature and supplemental fuel is required. The composition of the waste stream has to be analyzed in order to determine its heating value and this can take from a few minutes to a quarter hour. In addition, the time delays between the sample point and the analyzer and between the flow control valve position and the flare have to be considered in order to manage the heating value of the gases being flared appropriately. This presentation looks at the modeling and optimization of the flare system. The objectives are to reduce supplemental fuel use while maintaining heating value requirements.

402526 Advanced model development from production data using multiple objectives

Abstract:

The development of a model for a production reaction system has numerous challenges. The production process even in the best of situations will not include all measurements that the model will be predicting. There will be noise in the measurements, failure of sensors, and systematic errors. There will also be unmeasured factors that can impact the data. There will also be experimental data and designed data that needs to be incorporated into the model.

The model is usually designed to use some of the measurements as inputs while predicting the other measurements. A combination of transport theory and reaction theory is incorporated into the final model with a number of fitting parameters to correct for the unknowns and the simplifications which have been assumed. Often a weighting is assigned to different measured variables to designate their importance to the predictions. The different fitting parameters are adjusted to minimize the weighted difference between the results and the measurements. These fitted parameters can vary greatly depending on the weighting factors that are used.

In this presentation, we look at addressing the errors in the measurements and adjusting fitting parameters to match multiple results using a variety of optimization techniques. An understanding of the appropriate weightings between the different objective functions is gained by using Pareto front analysis. The fitting process is also used to provide insights into the measurement errors and the problems with some of the underlying assumptions used to build the original model. The model is then used to provide further insights into the production process.

402527 Discrete event simulation applied to rail fleet management

Abstract:

Commodity businesses in The Dow Chemical Company rely on large and expensive fleets of railcars to ship their products to market. In some cases a fleet can be shared by several sites and businesses. Managing such a fleet while balancing the level of empty and full rail cars at each site is a complex problem due to variations in demand, production, inventory targets, transit time, etc. A Discrete Event Simulation model was developed to represent the operation and management of a shared railcar fleet. This model was used to make decisions that will reduce freight, inventory, and railcar switching costs while maximizing the utilization of the fleet.

AUTHOR INDEX

AUTHOR INDEX

American Institute of Chemical Engineers
3 Park Avenue
New York, New York 10016-5991